SOILS IN CONSTRUCTION

Third Edition

W. L. Schroeder
OREGON STATE UNIVERSITY

JOHN WILEY & SONS
NEW YORK   CHICHESTER   BRISBANE   TORONTO   SINGAPORE

Library of Congress Cataloging in Publication Data:

Schroeder, W. L. (Warren Lee), 1939–
  Soils in construction

  Includes index.
    1. Soil mechanics.   2. Foundations.   3. Build.
I. Title.

TA710.S286   1984   624.1'5136   83-14569
ISBN 0-471-86581-8

Printed in the United States of America

10 9 8 7 6

To my parents, who taught me to work,
and my wife, who keeps me at it.

PREFACE

This book has been prepared as a teaching aid for a course in the Construction Program of the Civil Engineering Department at Oregon State University. The purpose of the course is to introduce students to the nature of soils and to illustrate how soil materials may influence certain construction operations. The course is not design oriented. It is a terminal course specifically arranged to deal with soils in construction for those who do not anticipate further study of soils and foundation engineering. This book is, therefore, suited for use in other, similar programs.

The book is divided into three parts. Part One is an introduction to soil materials. In conjunction with the testing methods in Part Three, the material presented provides the basic background for understanding soil behavior and how construction specifications relate to it. Part Two deals with soils in the construction contract. Contents of the usual contract and provisions regarding the influence of soil materials are discussed. Part Three is an appendix of testing materials. Today, persons from many fields interact with the design professional and they need to know and understand the designer's language to make use of this technology. It is to these persons that this textbook is directed.

The third edition manuscript was reviewed and criticized by Professor Theodore Marotta of Hudson Valley Community College, Professor Hank Mol of Auburn University, Professor Jack W. Martin of the University of Florida, Professor Gary L. McGavin of Riverside City College, Professor Thomas Foster of Texas Southern University, and Professor Gordan Larew of the University of Virginia. Pauline Eubanks and Laurie Campbell typed the manuscript and assisted with organization and indexing. I also wish to acknowledge the contribution made by reviewers of the second edition manuscript: Professor Glenn Staley of Southern Illinois University at Carbondale, Professor Thomas M. Dalton of the State University of New York at Canton, Professor James S. McKinney of Spartanburg Technical College, Professor P. V. Ramakrishnaiah of Texas Southern University, Professor Dell Orey Nickell of California Polytechnic State University, and Professor Carlton S. Yee of Humboldt State University.

I am indebted to Ray Layton and Professor Harold Ball for their helpful criticisms of the original drafts of the text. Larry Hansen assisted with the preparation of problems.

W. L. Schroeder
Corvallis, Oregon

CONTENTS

SOILS IN CONSTRUCTION

PART ONE

SOIL MATERIALS

CHAPTER 1

PHYSICAL CHARACTER OF SOIL CONSTITUENTS

Soil may be defined as an accumulation of solid particles produced by mechanical and chemical disintegration of rocks. It may contain organic constituents and water. This broad definition applies to a construction material that varies widely in its physical composition and behavior from location to location, and even on a particular site. This chapter describes the nature of various soil constituents and how they are developed from parent rock. Knowledge of the physical character of soil constituents is essential to an understanding of soil behavior during construction.

Studying this chapter should give the reader

1 A definition of soil material constituents and their origin.
2 Descriptive information concerning the size of soil particles of various types, and other factors that control soil behavior.
3 An introduction to the fundamental reasons for the observed differences in behavior among various fine-grained soils.

1.1 WEATHERING PROCESSES

Natural mechanical and abrasive forces degrade massive rock into smaller and smaller particles. These forces may be of thermal or gravitational origin. Mechanical weathering of parent rock produces coarser soil particles such as gravel and sand, and also silts which are finer. Soil particles produced by mechanical weathering have approximately three-dimensional shapes.

The less stable minerals in rock are chemically altered over long periods of time. Chemical weathering produces very small particles that are often in crystalline form. These particles are typically two-dimensional or flake shaped. Chemical weathering of rock results in clayey soils. The characteristics of clays depend on the nature of the parent rock, the environment in which the weathering takes place, and the length of time available for chemical alteration to develop.

1.2 CHARACTER OF THE COARSE SOIL FRACTION

Soil fractions may be designated by subdivision of the entire mass of solid material into size ranges. One such subdivision is listed in Table 1.1. The size ranges shown are arbitrary but represent those corresponding to the currently most popular nomenclature. Coarse soil constituents may be sand, gravel, or cobbles. Fine soil constituents consisting of silts and clays are discussed in the following section. An example of a laboratory sieve analysis used to separate soil into size fractions (ASTM D422) is shown in Figure 1.1.

The data shown in Figure 1.2 illustrate a test in which the particle size distribution, including percentage of fines, but not the fine particle size distribution, is to be determined. The procedure used is similar to ASTM D1140. Distribution of particle sizes finer than the No. 200 sieve is developed

Figure 1.1 Soil fractions separated by sieving. Each fraction shown is the range of particle sizes passing the next larger sieve and retained on the sieve with which it is shown.

SIEVE ANALYSIS DATA AND GRAIN SIZE CURVE

TEST FOR	Backfill for footings (7268)	
DATE	12 June 1972	
TEST BY	REF	
SAMPLE DESCRIPTION	Sandy gravel – collected from borrow pit	
	Sample No. 7268 (12)	

U.S. STANDARD SIEVE SIZE	WEIGHT RETAINED	TOTAL WEIGHT RETAINED	TOTAL WEIGHT PASSING	PERCENT FINER
3"	0	0	956.5 g	100.0
1½"	212.3 g	212.3 g	744.2 g	77.8
¾"	142.6 g	354.9 g	601.6 g	62.8
NO. 4	209.1 g	564.0 g	392.5 g	41.0
NO. 10	152.2 g	716.2 g	240.3 g	25.1
NO. 40	93.6 g	809.8 g	146.7 g	15.3
NO. 100	52.8 g	862.6 g	93.9 g	9.8
NO. 200	61.6 g	924.2 g	32.3 g	3.3
PAN TOTAL	32.3 g	956.5 g	0	0

FINE FRACTION	
UNWASHED WEIGHT	366.1 g
WASHED WEIGHT	360.2 g
FINES BY WASHING	5.9 g
FINES BY SIEVING	26.4 g
FINES TOTAL	32.3 g

NOTES

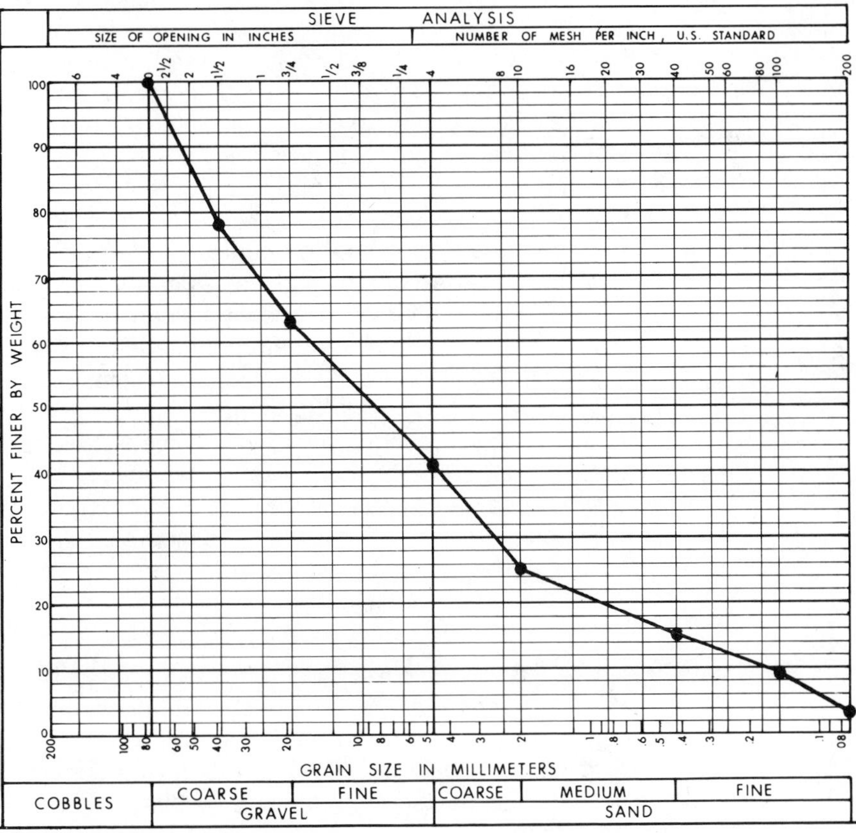

Figure 1.2 Data and results from laboratory sieve analysis.

TABLE 1.1   SOIL SIZE FRACTIONS

Constituent	U.S. Standard Sieve No.	Sieve Opening
Cobbles	Above 3 in.	3 in.
Gravel		
Coarse	3–¾ in.	3–¾ in.
Fine	¾ in.–No. 4	¾ in.–4.76 mm
Sand		
Coarse	No. 4–No. 10	4.76–2.00 mm
Medium	No. 10–No. 40	2.00–0.42 mm
Fine	No. 40–No. 200	0.42–0.074 mm
Silts and clays	Below No. 200	0.074 mm

from hydrometer test results (ASTM D422). The sample was first sieved dry. It was determined that 26.4 g of soil passed the No. 200 sieve. The washing and sieving procedure described in ASTM D1140 was then followed for that portion of the sample passing the No. 4 sieve. The tabulated weights retained are those determined by both dry sieving and washing. About 5.9 g of fines were removed from the minus No. 4 fraction by washing. The combined weight of fines in the sample, 32.3 g, is represented by the sum of the weights from sieving and washing. Weights retained on each sieve and the pan are tabulated. From these numbers, the total or cumulative weights retained on a given sieve are calculated. The total weight of the sample, less the total weight retained on any sieve, gives the total weight passing, which can then be expressed as a percentage.

Table 1.1 shows the coarse soil fraction subdivided according to size of individual particles. Other properties of the particles may be important in describing their acceptability for use as construction materials. Particle shape, or angularity, for instance, greatly influences the ability of a granular or coarse-grained soil to form interparticle interlocks. Degrees of angularity are illustrated in Figure 1.3. Individual particles may be angular (sharp distinct edges) or rounded. Subangular particles have distinct but rounded edges while subrounded particles are smooth and nonequidimensional. The greater the degree of sharpness and angularity of its particles, the greater ability a coarse soil has to form a stable, strong soil mass.

Specific gravity of a material is defined as

$$G_s = \frac{\gamma_S}{\gamma_W} \tag{1.1}$$

where γ_S is the unit weight of a material and γ_W is the unit weight of water at the reference temperature, 4 C. The specific gravity of soil constituents usually reported is an apparent specific gravity. Figure 1.4 indicates the distinction between apparent and absolute specific gravities. In determinations of specific gravity, the volume of solids and the weight of solids are determined in order that density may be computed. The normal procedure

Figure 1.3 Angularity variations for sand particles. *(a)* Platte River sand, Colorado (magnification 50×). *(b)* Ottawa sand (magnification 50×).

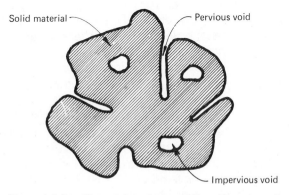

Figure 1.4 Specific gravity-volume relation. For apparent specific gravity, the particle volume includes solid material and impervious voids. For absolute specific gravity, the particle volume includes solid material only.

(ASTM D854) used to determine specific gravity involves volume determination by water displacement. Since impervious voids may not be filled by this process, the specific gravity normally reported is less than the absolute value. The importance of this distinction is usually minor. The specific gravity of most soil constituents is around 2.7. An example data sheet for laboratory specific gravity determinations is shown in Figure 1.5.

In addition to size and specific gravity, toughness and durability of coarse particles can be important factors. The quality of rock proposed for a specific use must often meet accepted standards for resistance to mechanical or chemical degradation. For these purposes the Los Angeles abrasion test (ASTM C131) and the Soundness of Aggregates by Use of Sodium Sulphate or Magnesium Sulphate (ASTM C88) test are widely used. In these tests, representative samples are subjected to harsh mechanical or chemical action and their weight loss is determined. Performance is compared with performance of materials that have proven acceptable when construction specifications are written. Other characteristics of the coarse soil fraction such as surface texture and surface chemistry may control their acceptability. These are most important when the material is used for aggregate in Portland cement or asphalt concrete.

1.3 CHARACTER OF THE FINE SOIL FRACTION

The fine soil fraction consists of silts and clays (see Table 1.1). Silts are usually similar in shape to coarse fraction particles. They are very small in size. Silt is the result of mechanical weathering. Clays result from chemical weathering. The difference in origins of silts and clays produces very distinct differences in their behavior.

Clays are typically hydrous aluminum silicates composed of distinct structural units with unique relationships to one another. The fundamental structural units are tetrahedral and octahedral ionic arrangements such as those shown in Figure 1.6. These units are assembled in sheets parallel to their basal planes. The silica units form a sheet with common oxygen ions at the corners. The basal oxygen ions share an electron with an adjacent silica. The resulting electrostatic charge per unit is -1 since, of the available -8 due to the oxygens, three are balanced by sharing with an adjacent silica and four are balanced by the central silica. The octahedral unit assembles with each hydroxyl common to three units. The resulting charge per unit is $+1$ since the available six negatives are divided by three.

The octahedral or tetrahedral sheets formed as described above are mutually attractive and unite in various ways to form the basic two-dimensional structure of clay minerals. The more common forms are shown schematically in Figure 1.7. These structures tend to exhibit excess negative charges on their flat sides, and positive charges at their edges.

SPECIFIC GRAVITY TEST

TEST FOR _Olequa Dam_

TEST BY _EFS_   DATE _4 March 1971_

SAMPLE DESCRIPTION _Red-brown clay core material._
test pit no. 1   depth - 12 feet

DETERMINATION NO.	1	2	3	4
BOTTLE NO.	5	5		
WT. BOTTLE + WATER + SOIL (W_1)	674.72	675.81		
TEST TEMPERATURE t, °C	22.8	23.2		
WT. BOTTLE + WATER (W_2)	643.13	643.09		
EVAPORATING DISH NO.	SM 2	SM 4		
WT. DISH + DRY SOIL	824.58	850.00		
WT. DISH	774.99	798.90		
WT. DRY SOIL (W_S)	49.59	51.10		
SPECIFIC GRAVITY OF WATER AT $t, (G_t)$	0.9976	0.9975		
SPECIFIC GRAVITY OF SOIL (G_S)				

REMARKS _____

_____

_____

$$G_S = \frac{G_t \, W_S}{W_S + W_2 - W_1}$$

1. $\dfrac{0.9976 \times 49.59}{49.59 + 643.13 - 674.72} = 2.72$

2. $\dfrac{0.9975 \times 51.10}{51.10 + 643.09 - 675.81} = 2.77$

Figure 1.5 Results and calculations of specific gravity test.

The structure of kaolinite is schematically represented by Figure 1.7*b*. Adjacent units tend to stack, one on the other (see Figure 1.8), as the oxygens in the basal layer of the tetrahedral unit are attracted to the hydrogens in the octahedral unit. A comparatively weak hydrogen bond results.

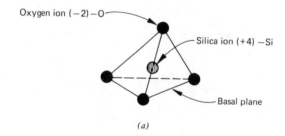

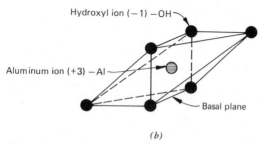

Figure 1.6 Fundamental clay building blocks. (After Grim, 1953.) *(a)* The silica tetrahedron. *(b)* The octahedral unit.

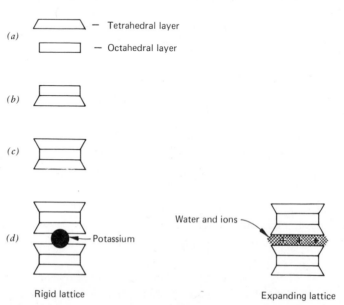

Figure 1.7 Schematic clay mineral structures. (After Scott, 1963.) *(a)* The basic units. *(b)* 1:1 clay mineral structure. *(c)* 2:1 clay mineral structure. *(d)* Typical 2:1 clay lattice structures.

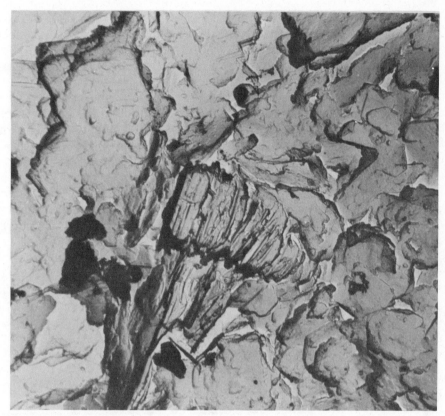

Figure 1.8 Parallel particle stacking in a clay soil—a ped (magnification 25,000 ×).

Kaolins are very stable minerals and are among the last formed in the weathering process. They exhibit very little physicochemical activity compared to other clay minerals, because of a comparatively large grain size and low surface charge.

Illite exhibits intermediate physicochemical activity. Interlayer potassium ions form a weak ionic bond with the oxygen ions in the silica sheets. A deficiency of potassium results in greater unsatisfied valence charge and correspondingly greater activity. Illite is a very commonly occurring clay mineral.

Montmorillonite is a third characteristic clay mineral. In the absence of the interlayer bonding noted in illite's structure the interlayer space is expandable. Typically it is filled with water. Depending on the abundance or deficiency of water available to occupy this space, a montmorillonitic soil may exhibit marked volume changes upon drying or wetting. These materials are particularly troublesome when abundant water is present and are the most claylike (active) of the clay minerals. They are among the poorest of available soils for most construction purposes. Montmorillonites are very unstable chemically and are among the initially formed chemical-weathering products. They derive principally from volcanic parent materials in a moist environment.

Other clay minerals have special properties that may assume major importance in construction works. Halloysite, for instance, is similar in structure to kaolinite, except that the plately particles tend to assume a tubelike shape. The shape is thought to result from the presence of interlayer adsorbed water, which weakens the hydrogen bond between the tetrahedral and octahedral sheets. A halloysite develops an irreversible change upon drying and becomes less claylike than in its natural state. Laboratory procedures that require drying before testing may, thus, misrepresent the true nature of a natural material, often to the detriment of the contractor. Drying in preparation for testing results in tests that may show that the soil is less plastic and, therefore, more workable than it actually may be in the field where drying has not occurred.

Amorphous (noncrystalline) materials occur widely. These soils are claylike yet do not have the simple structure described for crystalline clays. Very finely divided amorphous forms typically occur as gels and, in sufficient quantity, coat other mineral particles. In this manner, they serve to control soil behavior. While amorphous materials are an interesting phenomenon to some, to distinguish them from other clay soils may be, for the construction engineer, largely academic.

1.4 SUMMARY

This chapter has shown that soils occur as coarse-grained or fine-grained materials. Coarse soils are sands and gravels, and fine soils are silts and clays. If a soil particle, say a coarse sand grain, is subdivided, its volume remains the same while its surface area increases. Further subdivision of the resulting particles increases the area geometrically while total volume and weight are constant. The sand particle, if subdivided often enough, is eventually reduced to a large number of particles in the clay-size range. If the specific surface of a soil is defined as the area of the particle surfaces per unit of weight, an index of the effect of particle size on surface area is established. Consider the information in Table 1.2.

TABLE 1.2   SPECIFIC SURFACES OF SOIL MATERIALS

Material	Specific Surface (m^2/g)
Sand (0.1 mm)	0.03
Kaolinite	10
Illite	100
Montmorillonite	1000

Source: T. W. Lambe and R. V. Whitman, *Soil Mechanics*, Wiley, New York, 1969.

If it is recalled that the surfaces of clay minerals possess an electro-chemical charge, it is evident from Table 1.2 that the presence of this charge, as a factor in determining soil behavior, becomes increasingly important as particle size decreases. Many of the observed differences in soil behavior for coarse and fine materials may be explained on the basis of this infor-mation. Coarse soil behavior is controlled largely by the weight of the soil particles, while fine soils, clays in particular, owe much of their often trou-blesome nature to interparticle physicochemical forces that arise from their natural electrostatic charges.

The reader should now

1 Be able to distinguish among soil constituents on the basis of particle size.
2 Know how to explain the origin of surface charge on fine soil particles.

REFERENCES

American Society for Testing and Materials, *1974 Book of ASTM Stand-ards,* Philadelphia.
Grim, R. E., *Clay Mineralogy,* McGraw-Hill, New York, 1953.
Lambe, T. W., and Whitman, R. V., *Soil Mechanics,* Wiley, New York, 1969.
Scott, Ronald F., *Principles of Soil Mechanics,* Addison-Wesley, Reading, Mass., 1963.

CHAPTER 2

NATURAL SOIL DEPOSITS

Knowledge of the nature of soil constituents is of little value unless supplemented by information concerning the properties of assemblies of particles in a soil deposit. It has been said that to fully understand soil behavior it is essential that one have at least a basic understanding of geologic processes. Some of these concepts and their effect on the nature of natural soil deposits are discussed in this chapter. Additional study of the geologic sciences will prove valuable to engineering and construction students. The purposes of this chapter are

1 To introduce the concept of soil structure, and how individual soil particle arrangements arise.
2 To explain how differences in particle arrangements may affect mass soil behavior.
3 To distinguish between residual and transported soils.
4 To illustrate typical examples of various natural soil deposits.

2.1 SOIL STRUCTURE

The primary structure of a coarse soil is typically single grained. Individual particles may assume relatively stable or unstable positions according to their mode of deposition. The dense configuration shown in Figure 2.1a typically occurs in deposits built in an active water environment. Beach sands and river gravels are representative examples. In the absence of moving water or some other agent to affect the necessary arrangement, coarse soil deposits may be loose. Loose deposits are typically formed in quiet water or may result when a dense deposit is disturbed, by a landslide, for example.

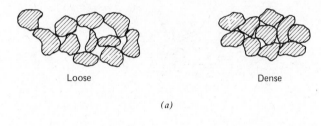

Loose   Dense

(a)

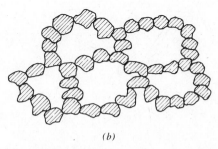

(b)

Figure 2.1 Some primary soil structures. (After Taylor, 1948.) (a) Single grain. (b) Honeycomb.

Very fine sands or silts may assume a honeycomb configuration similar to that in Figure 2.1b. In some instances the gravitational forces during deposition of such materials are not sufficient to overcome interparticle attractive forces, and a very open structure results. The term *metastable* is sometimes used to describe this condition because of its inherent sensitivity to even the most minor disturbance. Soil deposits of this nature often appear to be firm and strong but become wet and unworkable during the excavation process.

Clay soil exhibits a structure very strongly influenced by the chemical environment existing during deposition or by stress history thereafter. In the case of clay soils resulting from weathering of rock in place, the relic structure of the parent material may be evident in the resulting soil.

Clay soil particles deposited as sediment in freshwater repel one another as they settle because of the like electrostatic surface charge they carry. In the resulting bottom deposit the particles align themselves in a face-to-face arrangement as the repulsive forces from the surface charges are balanced. Forces due to the soil weight are resisted by electrostatic repulsion until an equilibrium condition similar to Figure 2.2a is reached. If the clay, on the other hand, is deposited in saltwater, the excess of available free ions present in the water effectively neutralizes the particles' surface charges. While suspended they are, thus, free to attract each other gravitationally and form flocs or aggregates of particles. The single grains in these flocs are hap-

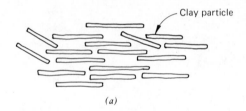

(a)

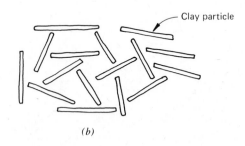

(b)

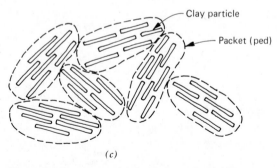

(c)

Figure 2.2 Structure or fabric of clay soils. *(a)* Dispersed. (After Lambe, 1958.) *(b)* Flocculated. (After Lambe, 1958.) *(c)* Packet or ped.

hazardly arranged as illustrated by Figure 2.2b. A very porous soil deposit results, with a structure that is characteristically weak and compressible.

Recently developed microscopy techniques have resulted in some new concepts of soil structure. It is now believed that some clays may exist in randomly arranged packets or peds, like those shown in Figures 1.8 and 2.2c, which individually are made up of highly oriented individual particles. This concept has been used to explain observed soil behavior inconsistent with previous concepts of structural arrangements.

Some soils develop a secondary structure during or after formation or deposition. A clay deposit, for example, may be highly dispersed by its own weight and subsequent drying as it is exposed to the atmosphere. As the drying continues or the surface is eroded a well-developed set of minute discontinuities is sometimes produced. Similarly, slickensides may be produced if a sediment is disturbed by landsliding or tectonic deformation. Discontinuities from any source may dominate the overall deposit's behavior.

2.2   SOIL DEPOSIT ORIGINS

Accumulations of soil particles form soil deposits in a well-defined sequence of geologic processes. Figure 2.3 illustrates the progression of mechanical and chemical weathering that converts all types of rock to residual soils, these soils' transport and deposition to form sediments, and the eventual conversion of sediments to sedimentary rock through a process of induration. While mechanical and chemical weathering begin anew, other geologic processes such as metamorphism and volcanic activity may alter the nature of the parent rock.

Typical residual soil profiles are shown in Figure 2.4. These soils develop in areas of stable topography through oxidation and leaching of the parent rock. The soil developed usually reflects the chemical and mineralogical makeup of the parent material. The soil layer may be further subdivided into *horizons* determined by their current chemical composition and state of weathering. Extensive residual soil deposits occur throughout the United States.

Where topographical features are unstable, we may have transported soils or no soils at all. A section through an area of deposition is shown in Figure

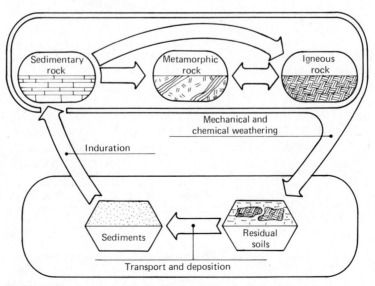

Figure 2.3 Geologic processes affecting origin of soil deposits.

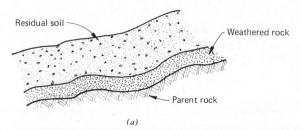

(a)

(b)

Figure 2.4 Residual soil profiles. *(a)* Schematic residual soil profile showing approximate parallelism in weathering zones. *(b)* Residual soil profile, derived from volcanic rock in western Oregon.

2.5*a*. A thin weathered zone of the underlying rock may or may not exist, depending on the geologic history of the region. Unlike the residual profile wherein the soil thickness is more or less constant, a transported soil deposit's thickness bears no relationship to bedrock contours whatsoever. The

TABLE 2.1   TRANSPORTING AGENTS AND SOIL DEPOSITS

Agent	Deposit Name	Depositional Environment
Water	Alluvium (Figure 2.6)	Flowing water
	Marine	Quiet brackish water
	Lacustrine	Quiet freshwater
Ice	Till (Figure 2.7)	Glacial ice contact zones
Wind	Loess (Figure 2.8)	Variable
	Dune sand	Arid or coastal lands
Gravity	Colluvium (Figure 2.9)	Below slide area
	Talus	Base of cliff

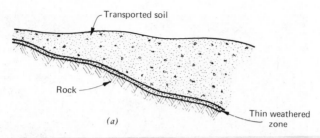

Transported soil

Rock

Thin weathered zone

(a)

(b)

Figure 2.5 Transported soil profiles. *(a)* Schematic transported soil profile. Note that soil depth bears no particular relationship to rock surface topography. The soil thickness varies, according to physical conditions during deposition. *(b)* Alluvial deposits at the mouth of the Madison River Canyon, Montana. Several levels of older terrace deposits are shown in the background.

degree of horizontal and vertical homogeneity within a transported deposit ranges from none to high, depending on the transporting agent and material source. Transported soils may be grouped according to the agents of transportation. Examples are given in Table 2.1.

Alluvial deposits may vary markedly both horizontally and vertically, owing to the nature of the current patterns in the parent stream. A lacustrine deposit may exhibit a *varved* secondary structure with horizontally uniform layers of fractions of an inch to several feet in thickness. The particle size within layers alternates from coarse to fine. This type of deposit forms in a lake when, during periods of high runoff, streams feeding the lake are able to carry large quantities of coarse sediment; then, during low flow periods only the finer particles are transported. Aeolian or wind-laid deposits are typically well sorted with very uniform particle size.

Figure 2.6 Cross-bedding in a natural sand deposit. Some transported soils, unlike this deposit, show a regular sequence of materials of uniform thickness.

Figure 2.7 Coarse glacial gravel deposit (till) in the State of Washington. Till may vary in particle size composition according to its source. Many midwestern U.S. tills consist principally of clays.

Figure 2.8 Excavation in loessial soil deposits near Vicksburg, Mississippi. True loess banks will normally stand with a vertical face because of cohesive strength resulting from natural cementation.

2.3 SUMMARY

The subsurface conditions in a natural deposit may often be anticipated with great accuracy if one understands the rudiments of the relations between landform and origin (geomorphology). Unanticipated subsurface conditions are one of the most, if not the most, frequent cause of disputes in execution of a construction contract. If all parties to the contract have adequate information regarding the nature of the subsurface materials and the sequence of their occurrence, and in addition are able to ascertain the effects these conditions will have on their respective operations, such disputes need not arise.

This chapter was intended to illustrate that the physical formation of a natural soil deposit may produce different soil structures. These different

Figure 2.9 Colluvial debris accumulated above residual soil profile in Oregon.

structural arrangements of soil particles result in observable differences in soil mass behavior. The reader should

1 Understand how and why these different structural arrangements come about.

2 Be able to identify and visualize different soil structures, and describe their behavior in general terms.

3 Know the difference between residual and transported soil deposit origins and be able to identify types of transported soils.

REFERENCES

Deere, D. U., and Patton, F. D., "Slope Stability in Residual Soils," *Proceedings, 4th Pan American Conference on Soil Mechanics and Foundation Engineering,* 1, 1971.

Lambe, T. W., "The Structure of Compacted Clay," *Journal, Soil Mechanics and Foundations Division, American Society of Civil Engineers,* 84, SM2, 1958.

Spangler, M. G., *Soil Engineering,* International Textbook Company, Scranton, Pa., 1969.

Taylor, D. W., *Fundamentals of Soil Mechanics,* Wiley, New York, 1948.

Terzaghi, K., and Peck, R. B., *Soil Mechanics in Engineering Practice,* Wiley, New York, 1967.

CHAPTER 3

SOIL INDEX PROPERTIES

In order to relate soil behavior to physical properties, it is convenient to have standard procedures for the testing and reporting of results. Test methods are used that give results indicative of the engineering behavior of soil materials. Some test results may permit judgments concerning the nature of individual soil particles, while others may be useful in accessing properties of an entire soil mass composed of individual particles. Generally speaking, the behavior of sands and gravels may be inferred from the shape, size, and density of packing of the constituent particles. The behavior of silts and clays is more nearly controlled by surface activity of the particles and in particular interaction of the particles with water.

This chapter introduces concepts and definitions for index properties of soil constituents and for the soil mass. The reader will learn about

1 The particle size distribution curve for coarse soils.
2 Plasticity characteristics for fine soils and their relationship to natural water content.
3 Phase relationships (air, water, and solid) for the soil mass.

3.1 COARSE-GRAINED CONSTITUENTS

Sands and gravels are most usefully described by reference to the distribution of particle sizes, the shape of particles, the relative density (see Chapter 8), and silt and clay content. In some cases the toughness and durability of particles is of major importance. All of these factors considered together determine the acceptability of a coarse soil for a particular use. Size distribution and fines content are usually represented on a semilogarithmic plot such as that shown in Figure 3.1. The distribution curves shown are determined by mechanical analysis as indicated in Chapter 1. The percentage of a sample greater than a given size is determined, for coarse

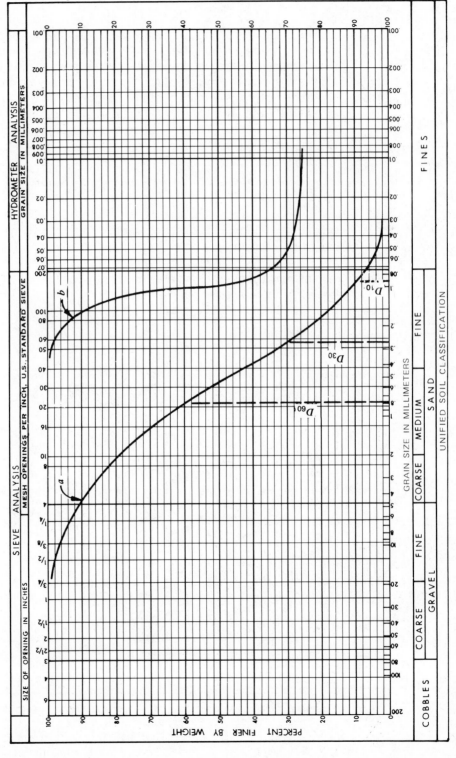

Figure 3.1 Particle size distribution curves.

soils, by passing the entire sample through a nest of standard sieves, and for fine soils, by hydrometer analysis (ASTM D422). The relation between percentage composition and size is thus determined. The maximum particle size in the finer nth percentage of a sample is indicated by the notation D_n. Curves a and b on Figure 3.1 represent two greatly different soil materials. The former contains a small percentage of fines and has a broad distribution of particle sizes. The latter is almost one-third silt and clay and two-thirds fine sand. It will be shown in Chapter 4 that indices from a particle size distribution curve are useful for soil identification and classification. The uniformity coefficient

$$C_u = \frac{D_{60}}{D_{10}} \qquad (3.1)$$

indicates the degree to which the particles are of the same size. If particles are all of the same diameter (see Figure 3.2a), C_u is 1. C_u for a real soil may, therefore, be any number greater than 1, the general rule being that increasing values represent an increasingly wide range of particle size differences. The coefficient of curvature

$$C_c = \frac{(D_{30})^2}{D_{60}D_{10}} \qquad (3.2)$$

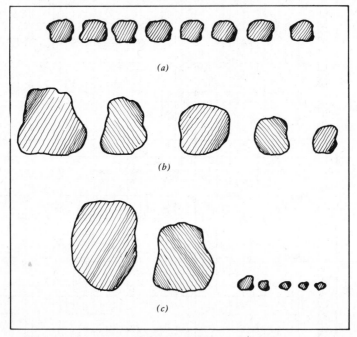

Figure 3.2 Particle size distributions. *(a)* Uniform size distribution. *(b)* Non-uniform size distribution. *(c)* Gap gradation.

describes the smoothness and shape of the gradation curve. Very high or very low values indicate that the curve is irregular.

3.2 FINE-GRAINED CONSTITUENTS

Because of the controlling importance of the effect of surface activity on behavior of fine-grained soils, description of these materials by reference to their particle sizes is practically meaningless. The practical distinction between silt and clay is made, not on the basis of an arbitrary size distinction, but on the basis of material behavior in the presence of water. Consider Figure 3.3. The consistency of fine soil varies according to the amount of water present. Completely dry, the soil may be hard (solid), while at high water contents it may be almost a slurry (liquid). Intermediate states of consistency are semisolid and plastic states. A plastic material is one that deforms readily without cracking or rupture. We define the boundaries of these states of consistency in terms of soil water content. The water content at the plastic-liquid boundary is the liquid limit (LL), while that at the plastic-semi-solid boundary is the plastic limit (PL). The difference between the liquid and plastic limits is the range of water contents over which a soil is plastic and it is designated as the plasticity index (PI). Different soils may be distinguished by their plasticity characteristics because these characteristics vary with surface activity of the constituent particles. The more active soils (claylike) are more plastic than the inactive soils (silts). This phenomenon may be explained by examining the nature of the water near the surface of a clay particle. Figure 3.4a is a conceptual view of a clay particle surrounded by water. Since water's molecular structure is dipolar, the water in the vicinity of the clay particle is effectively immobilized by the surface charge. It is adsorbed and may be considered essentially solid. As distance from the particle surface increases, the orientation of water is reduced in degree until, at the boundary of the particle's influence (limit of diffuse double layer), the viscosity is that of free water. With an abundance of water, soil particles would be separated by free water and the mixture would be fluid. As the amount of water decreases, the particles are separated by increasingly stiffer water. The mixture becomes like a solid. The soil water system,

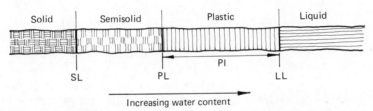

Figure 3.3 States of fine soil consistency. As water content increases, the soil becomes increasingly fluid.

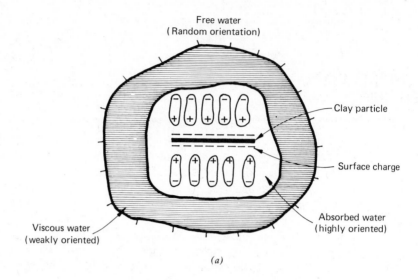

Free water
(Random orientation)

Clay particle

Surface charge

Absorbed water
(highly oriented)

Viscous water
(weakly oriented)

(a)

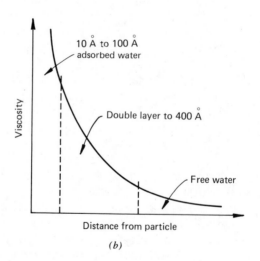

Viscosity

10 Å to 100 Å
adsorbed water

Double layer to 400 Å

Free water

Distance from particle

(b)

Figure 3.4 The clay-water system. *(a)* Conceptual clay-water system. *(b)* Viscosity in the double layer. (After Scott, 1963.)

fore, has a range of water contents over which it may be plastic. Changing the constituent phases (soil type or fluid) causes a change in the plasticity range. This permits us to distinguish among soils on the basis of their plasticity.

The plasticity index values are determined according to standard test procedures (ASTM D423 and ASTM D424), which are illustrated in Figure 3.5, and may be used to identify or classify fine soils. An example set of

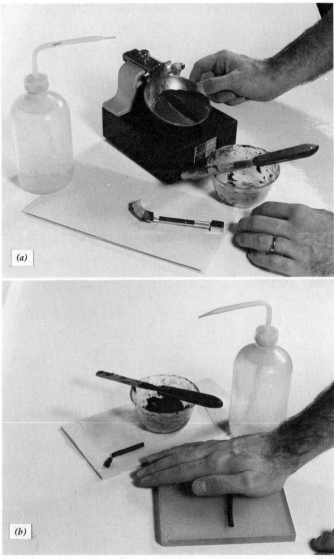

Figure 3.5 Atterberg limits tests for fine soil index properties. *(a)* The liquid limit test. *(b)* The plastic limit test.

test data is shown in Figure 3.6. The shrinkage limit (SL) may be determined (ASTM D427) but is not widely used as an index property. It represents the solid-semisolid interface and is the water content at which the soil volume is a minimum. High plasticity indices or high liquid limits are characteristic of clay soils. The use of these indices in identification and classification is described in Chapter 4.

ATTERBERG   LIMITS   AND   WATER   CONTENT

TESTS FOR *7253 - Murphy and Stephens Building*

TESTS BY *SHH*   DATE *28 September 1972*

SAMPLE DESCRIPTION   *ST·6, B·4*

PLASTIC   LIMIT   AND   WATER   CONTENT

TYPE OF TEST	W	PL	PL	PL		
CONTAINER NO.	14	16	17	18		
WET WT. + CONTAINER	16.21	4.40	3.85	5.70		
DRY WT. + CONTAINER	12.80	3.85	3.35	4.80		
WT. OF WATER	3.41	0.55	0.50	0.90		
WT. OF CONTAINER	1.35	1.60	1.40	1.45		
DRY WT. OF SOIL	11.45	2.25	1.95	3.35		
WATER CONTENT, %	29.78	24.44	25.64	26.86		

LIQUID   LIMIT

CONTAINER NO.	19	21	22	23		
NO. OF BLOWS	32	48	24	12		
WET WT. + CONTAINER	5.20	4.35	5.40	4.40		
DRY WT. + CONTAINER	4.20	3.55	4.25	3.50		
WT. OF WATER	1.00	0.80	1.15	0.90		
WT. OF CONTAINER	1.60	1.40	1.40	1.40		
DRY WT. OF SOIL	2.60	2.15	2.85	2.10		
WATER CONTENT, %	38.46	37.20	40.35	42.85		

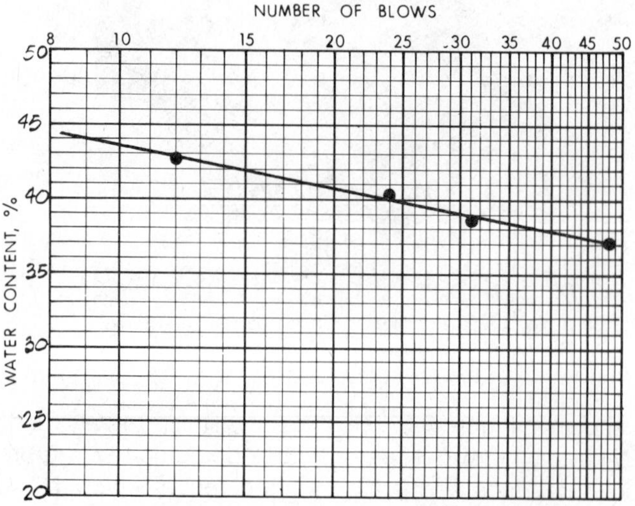

NUMBER OF BLOWS

W   *29.78 %*

LL   *40 %*

PL   *26 %*

PI   *14 %*

Figure 3.6 Data and calculations for water content and Atterberg limits test.

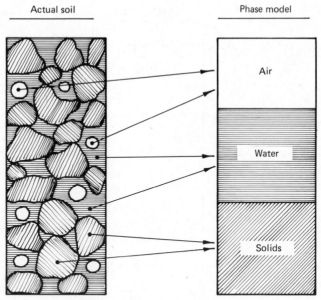

Actual soil   Phase model

Figure 3.7 Substitutions for phase calculations.

3.3 INDICES FOR THE SOIL MASS

Index values for coarse and fine constituents are descriptive of the physical nature of the constituents. They are not always sufficient to permit conclusions regarding soil performance for a given purpose. To supplement these indices we must also know soil density and water content and other relations among the air-water-solid phases of our material. Some of these relationships are defined and developed in the paragraphs that follow. It is convenient to substitute a conceptual model of the phases for the actual soil structure, as shown in Figure 3.7. Nomenclature applicable to the various constituent phases in a soil mass is shown in Figure 3.8. For the special cases of two-phase systems (either no water or all water in the voids), either V_W or V_A will be zero. Index values are defined in terms of this standard nomenclature. Some common examples are given in Table 3.1.

Consider a natural soil sample taken from a proposed borrow pit. The volume of the hole from which the sample was taken was 1.1 ft³. The total sample weighed 130 lb and, after drying, weighed 119 lb. The water content of the sample was thus

$$w = \frac{W_W}{W_S} \times 100 = \frac{11}{119} \times 100 = 9.2 \text{ percent}$$

The field density was

$$\gamma = \frac{W}{V} = \frac{130}{1.1} = 118 \text{ pcf}$$

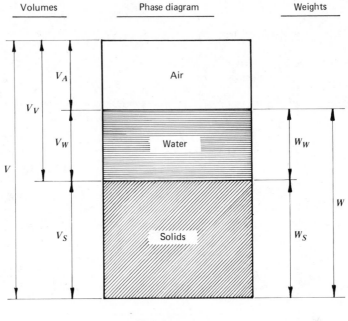

Volumes   Phase diagram   Weights

V_A Air

V_V

V_W Water W_W

V

V_S Solids W_S W

V = Total soil volume
V_V = Total void volume
V_A = Volume of air
V_W = Volume of water
V_S = Solid weight
W = Total weight
W_W = Weight of water
W_S = Weight of solids

Figure 3.8 Air-water-solid soil phases.

and the dry density was

$$\gamma_d = \frac{W_S}{V} = \frac{119}{1.1} = 108 \text{ pcf}$$

To determine some other index property it is convenient to represent what is known about the sample phases in a sketch similar to that of Figure 3.9. Since specific gravity (Equation 1.1) may be closely estimated (the actual

TABLE 3.1   SOIL MASS INDEX RELATIONS

Index Property	Symbol	Definition
Water content (express as percent)	w	(W_w/W_S)
Density (unit weight)	γ	(W/V)
Dry density (dry unit weight)	γ_d	(W_S/V)
Degree of saturation (express as percent)	S_R	(V_w/V_v)
Void ratio	e	(V_v/V_S)
Porosity (express as percent)	n	(V_v/V)

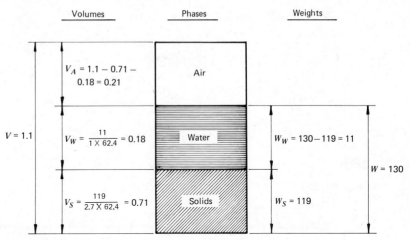

Figure 3.9 Example of air-water-solid phase calculations.

specific gravity could, of course, be determined by testing) the volume of the solid phase may be computed:

$$V_S = \frac{W_S}{G_S \gamma_w} = \frac{119}{2.7 \times 62.4} = 0.71 \text{ ft}^3$$

Similarly, the volume of water is

$$V_W = \frac{W_W}{G_W \gamma_w} = \frac{11}{1 \times 62.4} = 0.18 \text{ ft}^3$$

And by subtraction

$$V_A = V_V - V_W = V - V_S - V_W = 1.1 - 0.71 - 0.18 = 0.21 \text{ ft}^3$$

The weights and volumes of all phases are now known. Other defined indices may now be computed. For instance, the degree of saturation is

$$S_R = \frac{V_W}{V_V} \times 100 = \frac{0.18}{0.39} \times 100 = 46 \text{ percent}$$

It is often convenient to know the relationship between index properties rather than simply the definitions of these properties themselves. For instance, in discussing compaction of soils (Chapter 8), we must know how to compute dry density given wet density and water content. Density is defined as

$$\gamma = \frac{W}{V} = \frac{W_S + W_W}{V}$$

Also, dry density is

$$\gamma_d = \frac{W_S}{V}$$

Substituting for V,

$$\gamma = \frac{W_S + W_W}{V} = \gamma_d \frac{(W_S + W_W)}{W_S}$$

$$\gamma = \gamma_d(1 + w) \tag{3.3}$$

Or

$$\gamma_d = \frac{\gamma}{1 + w} \tag{3.4}$$

3.4 SUMMARY

The successful anticipation of soil behavior during construction may depend on availability of laboratory testing information for soils from the site. This chapter has presented certain indices of soil behavior that may be computed from laboratory test results. These same indices are often readily determined indicators of the engineering properties of the soil that have been presumed in design and must be obtained during construction. As such, index properties are specified as standards to meet in the execution of a contract. A thorough understanding of their significance is, therefore, essential.

Following study of this chapter and solution of the sample problems, the reader should

1 Be able to define soil constituent and soil mass index properties.
2 Know how to estimate consistency of fine soils given Atterberg limits and water content.
3 Thoroughly understand phase relation calculations.

REFERENCES

American Society for Testing and Materials, *1974 Book of ASTM Standards,* Philadelphia.

Lambe, T. W., and Whitman, R. V., *Soil Mechanics,* Wiley, New York, 1969.

Means, R. E., and Parcher, J. V., *Physical Properties of Soils,* Merrill, Columbus, Ohio, 1963.

Scott, Ronald F., *Principles of Soil Mechanics,* Addison-Wesley, Reading, Mass., 1963.

PROBLEMS

1 A 5-lb soil sample was taken from a fill. The volume of the hole from which the sample was taken was 0.044 ft^3. The sample weighed 4.4 lb after oven drying. It was determined that the specific gravity of the soil was 2.65. Compute:

   a. The total density of the fill.
   b. The dry density of the fill.
   c. The fill-void ratio.
   d. The fill porosity.
   e. The degree of saturation of the fill.
   f. The total density of the fill if its voids were to be filled with water (saturated density).

2 A saturated clay soil has a water content of 54 percent and a specific gravity of solids of 2.7. What are its total density and its dry density?

3 A stockpile of gravel that has a water content of 6 percent is rained on so that its water content increases to 9 percent. Compute the difference in the weights of dry gravel in a 10-ton truckload taken from the pile before and after the rain.

4. A sandy soil has a total density of 120 pcf, a specific gravity of solids of 2.64, and a water content of 16 percent. Compute:

   a. Its dry density.
   b. Its porosity.
   c. Its void ratio.
   d. Its degree of saturation.

5 If 1 ft^3 of soil with a void ratio of 0.9 is disturbed and then replaced at a void ratio of 1.16, what is its final volume? Express the final volume as a percentage of the original volume.

6 A moist soil has a total density of 110 pcf and a water content of 6 percent. Assuming that the dry density is to remain constant, how much water, in gallons per cubic yard, must be added to increase the water content to 12 percent?

7 A soil sample is obtained by pushing a 16-in.-long tube into the side of a pit. The tube has an inside diameter of 3 in. and weighs 1.8 lb. The soil in the tube is trimmed so that its ends are flush with the ends of the tube. If the total weight of the tube and sample is 8.62 lb and the water content of the soil is 43.6 percent, what are the density and dry density of the sample? If the soil is saturated, what is the average specific gravity of its solid constituents?

CHAPTER 4

SOIL CLASSIFICATION

Various systems have been devised for classifying soils such as those shown in Figure 4.1 in order to assign them descriptive names or symbols. Such systems are designed to group soils according to the physical characteristics of their particles or according to the performance they may exhibit when subjected to certain tests or conditions of service. Using classification symbols, it is hoped that engineers and contractors will be able to improve communications concerning site conditions. Soil classification systems described in this chapter are examples of systems available for engineering use and, in the case of the Unified and AASHTO systems, those most widely used.

The objectives of this chapter are to

1 Familiarize the reader with soil classification systems in general.
2 Show how popular soil classification systems are used.
3 Provide practice in the use of the Unified and AASHTO soil classification systems.

4.1 TEXTURAL CLASSIFICATION

Natural soils may be assigned descriptive names such as *silty clay, clayey sand,* and *sandy gravel* if the amounts of various constituent sizes are known. If the soil contains mostly sand and some clay, then clayey sand would seem appropriate. To establish such a system of classification one must assign size limits to the soil fractions and establish percentage compositions corresponding to descriptive names. The result will be a chart similar to that shown in Figure 4.2. This particular chart is the basis for soil descriptions published in soil maps by the U.S. Department of Agriculture. These maps,

Figure 4.1 A coarse soil and a fine soil. The coarse soil on the left is shown as it would appear wet or dry. The fine soil is shown in both states.

available for many counties in the United States, are sometimes useful references for engineering purposes, particularly where characteristics of only surficial soils are of principal concern.

On the chart of Figure 4.2, the index lines corresponding to the percent-

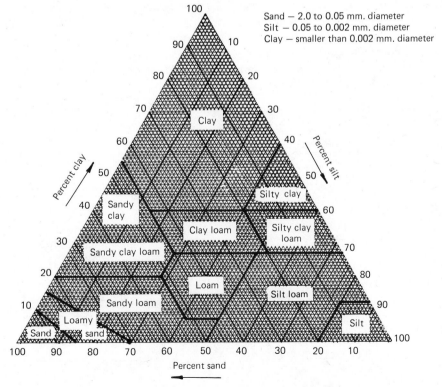

Figure 4.2 Textural soil classification chart, U.S. Department of Agriculture.

ages of each soil fraction are simply intersected. A soil with 50 percent clay, 20 percent sand, and 30 percent silt would be classified, for example, as *clay* since the intersection of the index lines is in the clay region. Index lines for clay are horizontal, index lines for silt are down and to the left, and index lines for sand are up and to the left, from the corresponding sides of the chart. If the soil contains 20 percent or more gravel, a *gravelly* prefix is added to the classification, and the percentages of sand, silt, and clay are taken as percentages of the soil fraction with the gravel excluded.

Textural classification of soils is of use to the builder primarily for coarse-grained soils (sands and gravels). It is these materials for which performance depends principally on the relative amounts and sizes of particles. For fine soils (silts and clays) textural classification provides little information of practical use for engineering purposes. The behavior of these soils is controlled by factors other than particle size. In particular, plasticity characteristics are of great importance.

4.2 UNIFIED SOIL CLASSIFICATION SYSTEM

The Unified Soil Classification System is used by the U.S. Bureau of Reclamation, the U.S. Corps of Engineers, the U.S. Forest Service, and many private consulting engineers. The system was developed for the Corps of Engineers during the 1940s. Its original purpose was to classify soils for use in roads and airfields, but it since has been revised to include use in em-

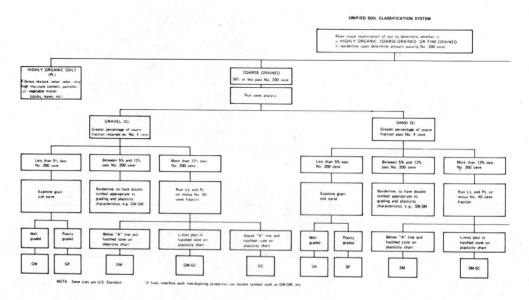

Figure 4.3 Flowchart for Unified Soil Classification System.

bankments and foundations as well. The Unified System permits classification of soils using either laboratory or field procedures. Field procedures are of particular value to the field engineer.

For a laboratory-based classification in the Unified System, index values of the soil are needed. If the soil is coarse (contains more than 50 percent sand and gravel), complete classification requires a grain-size distribution curve. From this curve, percentage composition of the constituents and the coefficients of uniformity and curvature are available. For the special case of a coarse soil containing more than 12 percent fines (silt and clay), the liquid limit and plasticity index of the fine fraction are also needed. With these indices the soil will be assigned a descriptive symbol consisting of two or more letters. These letters and their meanings are given in Table 4.1.

If the soil is fine grained, classification requires that we know both the liquid limit and plasticity index. The classification procedure for any soil is best illustrated by the flowchart in Figure 4.3 and the summary given in Figure 4.4. By beginning at the top of the chart and making decisions according to the criteria presented, the appropriate soil group symbol may be obtained. For fine soils, the distinction between silts and clays and high and low compressibility is based on the plasticity chart, Figure 4.5.

To illustrate tne use of these charts, assume that soil A with the gradation curve shown in Figure 4.6 is to be classified. It is first apparent that the soil is coarse grained. Only about 10 percent passes the No. 200 sieve. Also, only 10 percent of the soil (about 11 percent of the coarse fraction) is gravel.

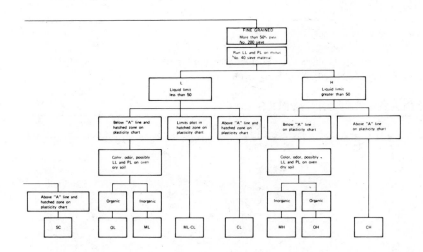

UNIFIED SOIL CLASSIFICATION

INCLUDING IDENTIFICATION AND DESCRIPTION

				FIELD IDENTIFICATION PROCEDURES (Excluding particles larger than 3 inches and basing fractions on estimated weights)	GROUP SYMBOLS (1)	TYPICAL NAMES
COARSE GRAINED SOILS — More than half of material is larger than No. 200 sieve size (2) (The smallest particle visible to the naked eye)	GRAVELS — More than half of coarse fraction is larger than No. 4 sieve size	CLEAN GRAVELS (Little or no fines)		Wide range in grain size and substantial amounts of all intermediate particle sizes.	GW	Well graded gravels, gravel-sand mixtures, little or no fines.
				Predominantly one size or a range of sizes with some intermediate sizes missing.	GP	Poorly graded gravels, gravel-sand mixtures, little or no fines.
		GRAVELS WITH FINES (Appreciable amount of fines)		Non-plastic fines (for identification procedures see ML below).	GM	Silty gravels, poorly graded gravel-sand-silt mixtures.
				Plastic fines (for identification procedures see CL below).	GC	Clayey gravels, poorly graded gravel-sand-clay mixtures.
	SANDS — More than half of coarse fraction is smaller than No. 4 sieve size (For visual classifications, the ¼" size may be used as equivalent to the No. 4 sieve size.)	CLEAN SANDS (Little or no fines)		Wide range in grain sizes and substantial amounts of all intermediate particle sizes.	SW	Well graded sands, gravelly sands, little or no fines.
				Predominantly one size or a range of sizes with some intermediate sizes missing.	SP	Poorly graded sands, gravelly sands, little or no fines.
		SANDS WITH FINES (Appreciable amount of fines)		Non-plastic fines (for identification procedures see ML below).	SM	Silty sands, poorly graded sand-silt mixtures.
				Plastic fines (for identification procedures see CL below).	SC	Clayey sands, poorly graded sand-clay mixtures

		IDENTIFICATION PROCEDURES ON FRACTION SMALLER THAN NO. 40 SIEVE SIZE				
			DRY STRENGTH (Crushing Characteristics)	DILATANCY (Reaction to Shaking)	TOUGHNESS (Consistency near Plastic Limit)		

			DRY STRENGTH	DILATANCY	TOUGHNESS	GROUP SYMBOLS	TYPICAL NAMES
FINE GRAINED SOILS — More than half of material is smaller than No. 200 sieve size (The No. 200 sieve size is about the smallest particle visible to the naked eye)	SILTS AND CLAYS — Liquid limit less than 50		None to slight	Quick to slow	None	ML	Inorganic silts and very fine sands, rock flour, silty or clayey fine sands with slight plasticity.
			Medium to high	None to very slow	Medium	CL	Inorganic clays of low to medium plasticity, gravelly clays, sandy clays, silty clays, lean clays.
			Slight to medium	Slow	Slight	OL	Organic silts and organic silt-clays of low plasticity
	SILTS AND CLAYS — Liquid limit greater than 50		Slight to medium	Slow to none	Slight to medium	MH	Inorganic silts, micaceous or diatomaceous fine sandy or silty soils, elastic silts.
			High to very high	None	High	CH	Inorganic clays of high plasticity, fat clays.
			Medium to high	None to very slow	Slight to medium	OH	Organic clays of medium to high plasticity.
HIGHLY ORGANIC SOILS			Readily identified by color, odor, spongy feel and frequently by fibrous texture.			Pt	Peat and other highly organic soils.

(1) Boundary classifications. Soils possessing characteristics of two groups are designated by combinations of group symbols. For example GW-GC, well graded gravel-sand mixture with clay binder.

(2) All sieve sizes on this chart are U.S. Standard.

Figure 4.4 Summary of Unified Soil Classification System. (Adapted from U.S. Department of the Interior, Bureau of Reclamation, *Earth Manual*, Denver, 1963.)

TABLE 4.1 UNIFIED CLASSIFICATION SYMBOLS

Symbol	Description
G	Gravel
S	Sand
M	Silt
C	Clay
O	Organic
Pt	Peat
W	Well graded
P	Poorly graded
L	Low compressibility
H	High compressibility
NP	Nonplastic

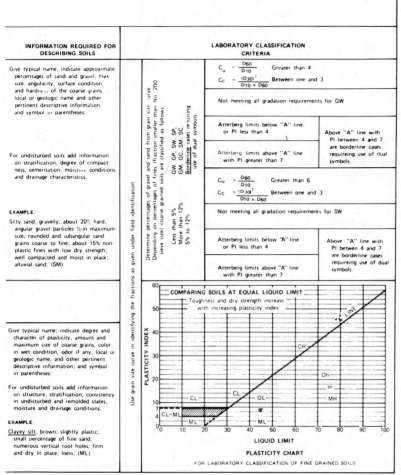

INFORMATION REQUIRED FOR DESCRIBING SOILS	LABORATORY CLASSIFICATION CRITERIA

Give typical name; indicate approximate percentages of sand and gravel, max. size, angularity, surface condition, and hardness of the coarse grains; local or geologic name and other pertinent descriptive information; and symbol in parentheses.

For undisturbed soils add information on stratification, degree of compactness, cementation, moisture conditions and drainage characteristics.

EXAMPLE:

Silty sand, gravelly; about 20% hard, angular gravel particles ½-in maximum size; rounded and subangular sand grains coarse to fine; about 15% nonplastic fines with low dry strength; well compacted and moist in place; alluvial sand; (SM)

Give typical name; indicate degree and character of plasticity; amount and maximum size of coarse grains; color in wet condition; odor if any; local or geologic name, and other pertinent descriptive information; and symbol in parentheses.

For undisturbed soils add information on structure, stratification, consistency in undisturbed and remolded states, moisture and drainage conditions.

EXAMPLE:

Clayey silt, brown; slightly plastic; small percentage of fine sand; numerous vertical root holes; firm and dry in place; loess; (ML)

Use grain size curve in identifying the fractions as given under field identification

Determine percentages of gravel and sand from grain size curve. Depending on percentages of fines (fraction smaller than No. 200 sieve size) coarse grained soils are classified as follows:
Less than 5% — GW, GP, SW, SP.
More than 12% — GM, GC, SM, SC.
5% to 12% — Borderline cases requiring use of dual symbols.

$C_u = \dfrac{D_{60}}{D_{10}}$ Greater than 4

$C_c = \dfrac{(D_{30})^2}{D_{10} \times D_{60}}$ Between one and 3

Not meeting all gradation requirements for GW

Atterberg limits below "A" line or PI less than 4 | Above "A" line with PI between 4 and 7 are borderline cases requiring use of dual symbols.

Atterberg limits above "A" line with PI greater than 7

$C_u = \dfrac{D_{60}}{D_{10}}$ Greater than 6

$C_c = \dfrac{(D_{30})^2}{D_{10} \times D_{60}}$ Between one and 3

Not meeting all gradation requirements for SW

Atterberg limits below "A" line or PI less than 4 | Above "A" line with PI betwen 4 and 7 are borderline cases requiring use of dual symbols.

Atterberg limits above "A" line with PI greater than 7.

COMPARING SOILS AT EQUAL LIQUID LIMIT
Toughness and dry strength increase with increasing plasticity index

PLASTICITY INDEX

"A" LINE

CH

OH or MH

CL

CL

OL

CL-ML

ML

ML

LIQUID LIMIT

PLASTICITY CHART
FOR LABORATORY CLASSIFICATION OF FINE GRAINED SOILS

ADOPTED BY – CORPS OF ENGINEERS AND BUREAU OF RECLAMATION
JANUARY 1952

The soil is, therefore, a sand. Since the fine fraction is between 5 and 12 percent of the sample weight, the soil will have a dual classification. By examining the gradation curve, we determine that

$$C_u = \frac{D_{60}}{D_{10}} = \frac{0.6 \text{ mm}}{0.074 \text{ mm}} = 8.1$$

and

$$C_c = \frac{(D_{30})^2}{D_{60} \times D_{10}} = \frac{(0.2)^2}{0.6 \times 0.074} = 0.9$$

According to Figure 4.4, the soil does not meet the requirements for a well-graded designation. It is, thus, a poorly graded sand. But the classification is not yet complete. Because of the high fines content, it must be determined

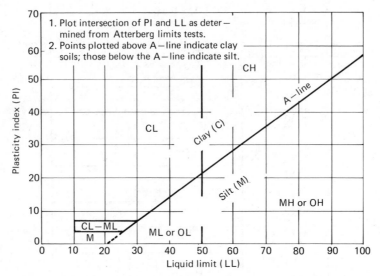

Figure 4.5 Plasticity chart for the Unified Soil Classification System.

if the fines are silty or clayey. If we know from laboratory tests that LL = 63 and PL = 42, we find that PI = 21. From Figure 4.5 it is determined that the fines are silty. The appropriate designation for the soil is thus SP-SM. In a similar manner, for soil B it is found that the appropriate designation is SP.

The previous example illustrated the use of the plasticity chart to classify fine-grained soils. The coordinates of liquid limit and plasticity index are plotted and the soil is assigned the symbol that represents the area in which the point falls. It should be noted that organic soils generally are distinguished from inorganic soils by color and smell. Dark soils with a distinctly objectionable smell (usually hydrogen sulfide) are organic. Positive identification of the presence of organics can be made in most cases by running liquid limit tests on the sample after air drying and again after oven drying. If the liquid limit is less than three-fourths the standard value because of oven drying, the soil is considered organic.

Field classification using the Unified System leads to the same results (classification symbols) as laboratory classification. The progress through the flowchart, Figure 4.3, is supplemented by simpler tests that are accomplished with little or no equipment. For fine soils, these tests are outlined in Figure 4.7. Responses typical of the various soil groups are also given and are further described in Figure 4.4.

For coarse soils the percentages of constituent sizes are simply estimated for field classification purposes. It is easiest to work by spreading the soil on a flat surface as shown in Figure 4.8. Coarse soils include particles that may be seen with the unaided eye. Individual fine soil particles cannot be seen without magnification. Therefore, a fine soil is identified in the field as

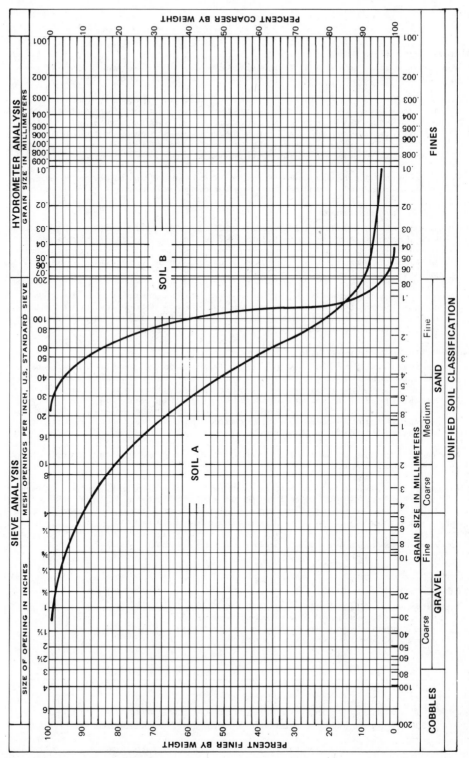

Figure 4.6 Gradation curves for soil classification examples.

Dilatancy (Reaction to shaking)

After removing particles larger than No. 40 sieve size, prepare a pat of moist soil with a volume of about one-half cubic inch. Add enough water if necessary to make the soil soft but not sticky.

Place the pat in the open palm of one hand and shake horizontally, striking vigorously against the other hand several times. A positive reaction consists of the appearance of water on the surface of the pat which changes to a livery consistency and becomes glossy. When the sample is squeezed between the fingers, the water and gloss disappear from the surface, the pat stiffens, and finally it cracks or crumbles. The rapidity of apperance of water during shaking and of its disappearance during squeezing assist in identifying the character of the fines in a soil.

Very fine clean sands give the quickest and most distinct reaction whereas a plastic clay has no reaction. Inorganic silts, such as a typical rock flour, show a moderately quick reaction.

Dry Strength (Crushing characteristics)

After removing particles larger than No. 40 sieve size, mold a pat of soil to the consistency of putty, adding water if necessary. Allow the pat to dry completely by oven, sun, or air drying, and then test its strength by breaking and crumbling between the fingers. This strength is a measure of the character and quantity of the colloidal fraction contained in the soil. The dry strength increases with increasing plasticity.

High dry strength is characteristic for clays of the CH group. A typical inorganic silt possesses only very slight dry strength. Silty fine sands and silts have about the same slight dry strength, but can be distinguished by the feel when powdering the dried specimen. Fine sand feels gritty whereas a typical silt has the smooth feel of flour.

Toughness (Consistency near plastic limit)

After removing particles larger than the No. 40 sieve size, a specimen of soil about one-half inch cube in size is molded to the consistency of putty. If too dry, water must be added and if sticky, the specimen should be spread out in a thin layer and allowed to lose some moisture by evaporation. Then the specimen is rolled out by hand on a smooth surface or between the palms into a thread about one-eighth inch in diameter. The thread is then folded and rerolled repeatedly. During this manipulation the moisture content is gradually reduced and the specimen stiffens, finally loses its plasticity, and crumbles when the plastic limit is reached.

After the thread crumbles, the pieces should be lumped together and a slight kneading action continued until the lump crumbles.

The tougher the thread near the plastic limit and the stiffer the lump when it finally crumbles, the more potent is the colloidal clay fraction in the soil. Weakness of the thread at the plastic limit and quick loss of coherence of the lump below the plastic limit indicate either inorganic clay of low plasticity, or materials such as kaolin-type clays and organic clays which occur below the A-line.

Highly organic clays have a very weak and spongy feel at the plastic limit.

NOTE:

These procedures are to be performed on the minus No. 40 sieve size particles, approximately 1/64 in. For field classification purposes, screening is not intended; simply remove by hand the coarse particles that interfere with the tests.

Figure 4.7 Field identification procedures for fine-grained soils or fine soil fractions. (Adapted from U.S. Department of the Interior, Bureau of Reclamation, *Earth Manual*, Denver, 1963.)

Figure 4.8 Visualizing the particle size distribution in a coarse soil sample.

one that contains more than 50 percent by weight of visually indistinguishable particles. It is important to point out that often aggregates of fine particles are mistaken for sand or gravel. Care should be taken in field classification to ensure that this difficulty does not arise. Having decided if a soil is coarse or fine, one then proceeds to complete the classification, as described in the following paragraphs.

For coarse soils the distinction between sand and gravel is based on passage through the No. 4 sieve. For field classification ¼-in. particle size is generally used. For clean sands or gravels (those with less than 12 percent fines) the gradation designation W or P is assigned depending on whether there is a wide range of particle sizes represented or if only one or two sizes predominate. Dirty sands and gravels have a second designator M or C, according to whether the fines are silty or clayey. This distinction is based on tests that are also appropriate for fine-grained soils.

In Figure 4.7 the dilatancy, dry strength, and toughness tests for fine soils are described. The reactions indicative of the various soil classifications are shown in Figure 4.4. Generally the distinction between silts and clays is easily made on the basis of dilatancy (ability to change volume) or dry strength, as shown in Figures 4.9 and 4.10. For moist soils the toughness test, described in Figure 4.7, is useful. It is more difficult, especially for silt, to distinguish between high and low compressibility. Probably the best method

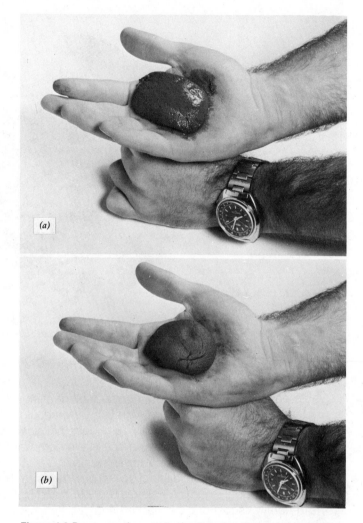

Figure 4.9 Response of a nonplastic silt to the dilatancy test. *(a)* After jarring the sample by tapping the hand to densify the sample (note shiny surface). *(b)* After opening the hand to allow the sample to expand (note dull surface).

of learning to do so is to observe the reactions to these simple tests of soils that have been classified using laboratory procedures. It is also useful, especially for beginners, to summarize observations on a data sheet, such as that shown in Figure 4.11, before assigning a classification.

Figure 4.10 Demonstration of dry strength of a clay (left) and a silt (right).

4.3 AASHTO CLASSIFICATION SYSTEM

The Unified Classification System groups soils according to their behavior in a wide variety of uses. The American Association of State Highway and Transportation Officials (AASHTO) and the Federal Highway Administration (FHWA) use a classification system intended to indicate behavior of materials used as highway subgrades. There are seven basic soil groups designated A-1 to A-7. Soils are placed in these groups according to performance characteristics, A-1 being the best and A-7 being the worst. An A-8 designation is reserved for peat or highly organic soils. A-1 to A-3 soils are sands and gravels, whereas A-4 to A-7 soils are silts and clays.

Some basic differences between the AASHTO System and the Unified System should be noted. For instance, gravel in the AASHTO System is that material passing the 3-in. sieve and retained on the No. 10 sieve. In the Unified System, the lower particle size bound is defined by the No. 4 sieve. The AASHTO System recognizes only fine and coarse sand. Fine sand has the same particle size range in both classification systems, while coarse sand in the AASHTO System is that material passing the No. 10 sieve and retained on the No. 40. Coarse sand in the AASHTO System is thus the same as medium sand in the Unified System. The smallest gravel-size ma-

VISUAL SOIL IDENTIFICATION

TEST BY _TKN_

FOR _7412 - Capital Corporation_   DATE _2 February 1974_

SAMPLE NO.	MAX. SIZE	ESTIMATED GRADATION			COLOR (Wet)	DESCRIPTION AND IDENTIFICATION — Consider particle size and shape, gradation, fines, plasticity, dry strength, reaction to shaking test, odor, undisturbed consistency.	GROUP SYMBOL
		GRAVEL (%)	SAND (%)	FINES (%)			
B6, ST 1	No. 40	0	10	90	dark brown	dark brown inorganic stiff silty clay	CL
B6, ST 2	No. 40	0	10	90	light brown	light brown inorganic soft clayey silt	MH
B6, ST 3	No. 40	0	10	90	light brown	light brown inorganic soft clayey silt	MH
B6, SS 1	No. 4	0	75	25	grey-brown	grey brown angular silty sand	SM
B6, SS 2	No. 4	0	75	25	grey-brown	grey brown angular silty sand	SM

Figure 4.11 Summary of observations for visual soil identification and classification.

General Classification	GRANULAR MATERIALS (35 percent or less of total sample passing No. 200)							SILT-CLAY MATERIALS (More than 35 percent of total sample passing No. 200)			
	A-1		A-3	A-2				A-4	A-5	A-6	A-7
Group Classification	A-1-a	A-1-b		A-2-4	A-2-5	A-2-6	A-2-7				A-7-5, A-7-6
Sieve analysis, percent passing:											
No. 10	50 max.										
No. 40	30 max.	50 max.	51 min.								
No. 200	15 max.	25 max.	10 max.	35 max.	35 max.	35 max.	35 max.	36 min.	36 min.	36 min.	36 min.
Characteristics of fraction passing No. 40:											
Liquid limit				40 max.	41 min.	40 max.	41 min.	40 max.	41 min.	40 max.	41 min.
Plasticity index	6 max.		NP	10 max.	10 max.	11 min.	11 min.	10 max.	10 max.	11 min.	11 min.*
Group Index**	0		0	0	0	4 max.		8 max.	12 max.	16 max.	20 max.

Classification procedure: With required test data available, proceed from left to right on chart; correct group will be found by process of elimination. The first group from the left into which the test data will fit is the correct classification.

*P.I. of A-7-5 subgroup is equal to or less than L.L. minus 30. P.I. of A-7-6 subgroup is greater than L.L. minus 30 (see Fig. 4.13)

**See group index formula or Fig. 4.14 for method of calculation. Group index should be shown in parentheses after group symbol as: A-2-6(3), A-4(5), A-6(12), A-7-5(17), etc.

Figure 4.12 AASHTO classification for highway subgrade material. (Courtesy, American Association of State Highway and Transportation Officials.)

terial in the AASHTO System is included as coarse sand in the Unified System.

An additional noteworthy difference between the two systems concerns the definition of fine and coarse soils. In the AASHTO System, fine soils (A-4 to A-7) are those containing 35 percent or greater fines. The corresponding fines content in the Unified System is 50 percent.

Material quality within the basic soil groups is further indicated by an additional number or letter designation, according to the results of sieve analysis and Atterberg limits tests. A summary of the classification system is given in Figure 4.12. The chart in Figure 4.12 is used by proceeding from left to right with the use of the available test data. The first soil group encountered in this manner that matches the test data is the correct classification. The plasticity chart shown in Figure 4.13 is of considerable aid in

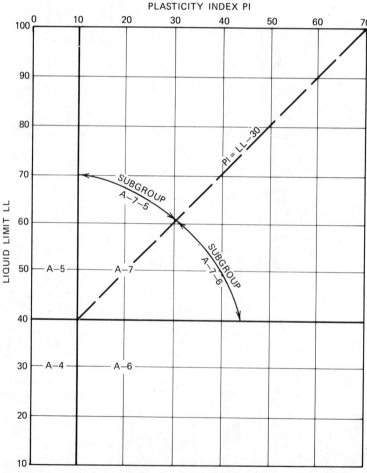

Figure 4.13 Plasticity chart for AASHTO soil classification. (Courtesy, American Association of State Highway and Transportation Officials.)

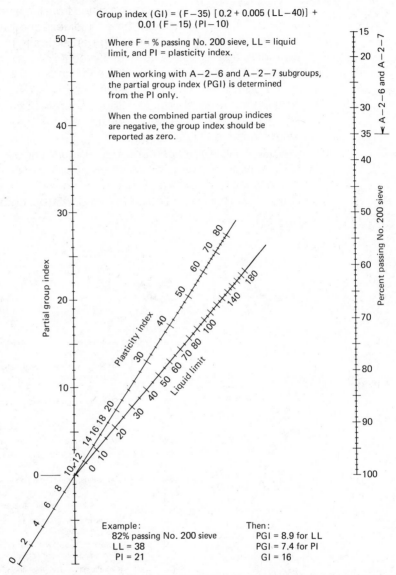

Group index (GI) = (F−35) [0.2 + 0.005 (LL−40)] +
0.01 (F−15) (PI−10)

Where F = % passing No. 200 sieve, LL = liquid
limit, and PI = plasticity index.

When working with A−2−6 and A−2−7 subgroups,
the partial group index (PGI) is determined
from the PI only.

When the combined partial group indices
are negative, the group index should be
reported as zero.

Example:
82% passing No. 200 sieve
LL = 38
PI = 21

Then:
PGI = 8.9 for LL
PGI = 7.4 for PI
GI = 16

Figure 4.14 Group index chart for AASHTO Classification System. (Courtesy, American
Association of State Highway and Transportation Officials.)

the selection of a classification for fine soils. It incorporates the same information as that shown in Figure 4.12 and may be used in place of it.

Because soil behavior is influenced not only by the amount of fine material present but also by its plasticity, AASHTO classifications include a parenthesized number that supplements the basic group symbols. As with the group symbols, increasing size of number indicates decreasing quality of material. The group index is determined from the nomograph shown in Figure 4.14, by adding two partial group indices, one based on plasticity index and one based on liquid limit. The partial group indices are determined by extending a straight line from the percent fines scale through the liquid limit scale and the plasticity index scale to intersect the partial group index axis.

The use of the AASHTO System may be illustrated by reclassifying the soils in the previous section. Soil A (Figure 4.6) classifies as an A-2-7 while soil B is an A-2-4. Soil A is considered a poorer material because of its moderately high fines content.

A sandy silt with 60 percent fines, a liquid limit of 67, and a plasticity index of 15 would classify as an MH in the Unified System and as an A-7-5 (11) in the AASHTO System. The group index (11) consists of partial group indices 8.5 (based on liquid limit) and 2.5 (based on plasticity index).

Major Divisions		Letter[a]	Name
(1)	(2)	(3)	(4)
COARSE-GRAINED SOILS	GRAVEL AND GRAVELLY SOILS	GW	Well-graded gravels or gravel-sand mixture, little or no fines
		GP	Poorly graded gravels or gravel-sand mixtures, little or no fines
		GM d u	Silty gravels, gravel-sand-silt mixtures
		GC	Clayey gravels, gravel-sand-clay mixtures
	SAND AND SANDY SOILS	SW	Well-graded sands or gravelly sands, little or no fines
		SP	Poorly graded sands or gravelly sands, little or no fines
		SM d u	Silty sands, sand-silt mixtures
		SC	Clayey sands, sand-clay mixtures
FINE-GRAINED SOILS	SILTS AND CLAYS LL IS LESS THAN 50	ML	Inorganic silts and very fine sands, rock flour, silty or clayey fine sands or clayey silts with slight plasticity
		CL	Inorganic clays of low to medium plasticity, gravelly clays, sandy clays, silty clays, lean clays
		OL	Organic silts and organic silt-clays of low plasticity
	SILTS AND CLAYS LL IS 50 OR GREATER	MH	Inorganic silts, micaceous or diatomaceous fine sandy or silty soils, elastic silts
		CH	Inorganic clays to high plasticity, fat clays
		OH	Organic clays of medium to high plasticity, organic silts
HIGHLY ORGANIC SOILS		Pt	Peat and other highly organic soils

Figure 4.15 Characteristics of Unified Classification soil groups relative to suitability for road and airfield pavement subgrades. (U.S. Army Corps of Engineers.)

4.4 SUMMARY

Soil classification is an aid in predetermining the behavior of soils under various conditions of service and during construction. The chart shown in Figure 4.15 is one example of how classification data may be used to characterize soil suitability for a specific purpose on the basis of limited and easily obtainable data. Such characterizations can provide very useful preliminary information on a job site. Classifications in the systems discussed are often found in the subsurface information section contained in contract documents for earthwork. It helps the bidder and the contractor during construction to understand the meaning of these various terms and symbols. They are particularly useful when supplemented by other information from investigative reports.

After having studied this chapter and solving the problems at the end, the reader should

Value as Subgrade When Not Subject to Frost Action (5)	Potential Frost Action (6)	Compressibility and Expansion (7)	Drainage Characteristics (8)
Excellent	None to very slight	Almost none	Excellent
Good to excellent	None to very slight	Almost none	Excellent
Good to excellent	Slight to medium	Very slight	Fair to poor
Good	Slight to medium	Slight	Poor to practically impervious
Good	Slight to medium	Slight	Poor to practically impervious
Good	None to very slight	Almost none	Excellent
Fair to good	None to very slight	Almost none	Excellent
Fair to good	Slight to high	Very slight	Fair to poor
Fair	Slight to high	Slight to medium	Poor to practically impervious
Poor to fair	Slight to high	Slight to medium	Poor to practically impervious
Poor to fair	Medium to very high	Slight to medium	Fair to poor
Poor to fair	Medium to high	Medium	Practically impervious
Poor	Medium to high	Medium to high	Poor
Poor	Medium to very high	High	Fair to poor
Poor to fair	Medium	High	Practically impervious
Poor to very poor	Medium	High	Practically impervious
Not suitable	Slight	Very high	Fair to poor

[a] (3) Subdivision of GM and SM groups is for road and airfield materials. Designation "d" applied to soils with LL ≤ 25 and PI ≤ 5. Designation "u" to others.

1 Have a reading knowledge of available engineering soil classification systems.

2 Be able to correctly classify soils in the AASHTO and Unified systems, given appropriate laboratory test data.

3 Be able to classify soils in the Unified System given only the soil to work with.

4 Know how to describe soils classified in the AASHTO System and the Unified, System given soil group symbols.

REFERENCES

American Society for Testing and Materials, *1974 Book of ASTM Standards,* Philadelphia.

American Association of State Highway and Transportation Officials, *Highway Materials, Part I, Specifications,* Washington, D.C., 1978.

Casagrande, A., "Classification and Identification of Soils," *Transactions, American Society of Civil Engineers,* **113,** 1948.

U.S. Department of Agriculture, *Soil Survey Manual,* Handbook No. 18, 1951.

U.S. Department of the Interior, Bureau of Reclamation, *Earth Manual,* Denver, 1963.

PROBLEMS

1 A dark gray soil is described as having a rapid reaction to the shaking or dilatancy test, no dry strength, and an offensive odor. Describe the soil and assign it a tentative classification in the Unified Soil Classification System.

2 Given the test results that follow, classify the soil in the Unified Classification System. Outline the steps you take in arriving at your result.

U.S. Standard Sieve No.	Percent Finer
4	100
10	93
40	71
200	48

Atterberg Limits of Minus No. 40 Fraction
LL = 33
PL = 30

3 Test results for a number of soils are shown in the following table. Classify each soil in the Unified System and in the AASHTO System, using your best judgment when specific information you might require is not given. Show any calculations you may find necessary and list any assumptions you make.

Soil No.	1	2	3	4	5	6	7	8	9
Sieve No.					Percent Finer				
4	73								67
10	50	92	100		86			88	43
40	40	52	69	100	61	92		78	19
100	25	41	52		30		100	19	16
200	15	28	46	62	19	90	69	7	7
LL	20	29	37	72	39	38	NP	NP	42
PL	15	15	30	45	30	19	NP	NP	21

4 Describe a soil classified as SW and a soil classified as CH in the Unified Soil Classification System as you would to a person having no knowledge of soil classification procedures or equipment.

5 Prepare a flowchart similar to that shown in Figure 4.3, for use in field identification of soil materials.

CHAPTER 5

STRESS ANALYSIS AND ENGINEERING PROPERTIES

The construction contractor is concerned with analyses using engineering properties of soils in those instances where construction of temporary work such as shoring for an excavation, falsework for a structure, or a dewatering system requires some engineering design using soil parameters. In some circumstances the contractor's staff does the analysis and design work. In other cases outside consultants are employed. In either event, the design requires some knowledge of engineering soil properties and the principles governing soil behavior. This chapter is an introduction to the subject. After studying this chapter, the reader should have developed an understanding of the effective stress concept and learned the principles of analysis for gravity stresses in soils. The engineering properties, soil strength, permeability, and compressibility will be related to soil index properties discussed earlier. It will be shown, for instance, that strength varies with the changing water content of clay soils and that permeability varies with the grain size of coarse soils. The reader should learn

1 The effective stress principle.
2 How to make static stress calculations.
3 About capillarity.
4 The relative permeabilities of various soils and how permeability values may be determined.
5 What is meant by compressibility.
6 The mechanism leading to consolidation of soils.
7 The factors controlling soil strength.
8 Relative strengths of troublesome soils and how they may be inferred from index properties.

5.1 THE EFFECTIVE STRESS PRINCIPLE

Soil engineering properties and behavior are strongly influenced by stresses and stress history. For this reason, it is important to understand the principles on which stress determinations are based, and how to make fundamental stress calculations.

Figure 5.1a shows a sphere of weight W_s and volume V_s resting inside a container with a square base of width b. If we assume that the container is weightless, then the average stress on the surface on which the container rests is W_s/b^2. This is the stress due to the weight of the sphere. The sphere contacts the inside of the container at a point, but we may think of its weight being distributed over the inner container base for purposes of illustration. The average stress on the base in these circumstances would be W_s/b^2, the same as on the supporting surface. If, as in Figure 5.1b, we fill the container with water to depth z, the stress on the base becomes

$$p = \frac{W_s + (zb^2\gamma_w - V_s\gamma_w)}{b^2} \tag{5.1}$$

We may think of Equation 5.1 as representing the total stress due to the weight of the sphere and water. The stress on the inner container base has two components. That due to water is obviously

$$u = z\gamma_w \tag{5.2}$$

By subtracting Equation 5.2 from Equation 5.1, we obtain the stress due to the sphere:

$$p' = \frac{W_s - V_s\gamma_w}{b^2} \tag{5.3}$$

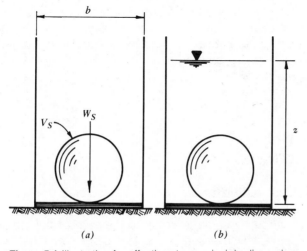

(a)   *(b)*

Figure 5.1 Illustration for effective stress-principle discussion.

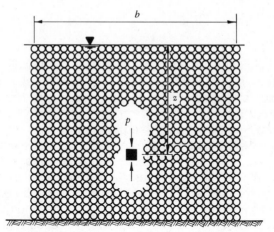

Figure 5.2 Idealized soil stratum.

The difference between the stresses due to the sphere in Figure 5.1*a* and 5.1*b* results from the buoyancy effect of the water. Implicit in the above discussion is the fact that

$$p = p' + u \qquad (5.4)$$

To illustrate how this observation is relative to soils, we may consider many spheres, representing soil particles, as shown in Figure 5.2. At some depth *z*, the average total stress due to all of the spheres and the water will be the sum of their respective weights divided by the area they occupy. In fact, if the sum of the weights is W_s and the sum of the volumes is V_s, the total stress for an area *b* × *b* is represented by Equation 5.1. Following the same logic as in the foregoing illustration, it is obvious that Equation 5.4 remains valid. This equation represents the effective stress principle, a most important principle in describing soil behavior in engineering applications. It may be stated as follows: *Total stress is equal to the sum of effective stress and porewater pressure* or, alternatively, *effective stress is equal to total stress minus porewater pressure.* Porewater pressure is also sometimes referred to as neutral stress.

5.2 VERTICAL EARTH PRESSURE CALCULATIONS

Vertical earth pressures may be calculated using Equation 5.4 and appropriate material properties. The total stress or pressure *p* at a point beneath the ground surface is equal to the sum of the water pressure *u* and the effective soil pressure *p'* at that point. The effective soil stress may be otherwise described as the sum of the forces at soil grain contacts within

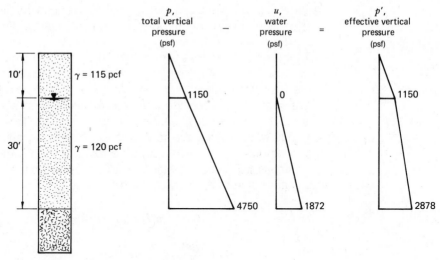

Figure 5.3 Example of vertical earth pressure analysis.

an area, divided by that area. A convenient way to calculate vertical pressure distribution with depth is illustrated in Figure 5.3. A simple stratigraphic column is shown and the unit weights of the soil materials involved are given. This column is representative of a 40-ft sand layer overlying rock. The water table is at a depth of 10 ft. The unit weight of the upper 10 ft may be considered as the total unit weight of the material. The unit weight shown for the lower 30 ft is also the total unit weight of the material but, in this case, the material is saturated. The total vertical pressure at any point is equal to the sum of the weights above a given area divided by that area as described in Section 5.1. If we consider 1 ft², then the vertical pressure at a point will be equal to the total unit weight of the material above that point times the vertical distance to the point. At 10 ft in depth, therefore, the total vertical pressure is 1150 psf. At 40 ft in depth, the total vertical pressure is the sum of the vertical pressure at 10 ft and the change in vertical pressure for the lower 30 ft, or 4750 psf. The water pressure at that depth is the product of the unit weight of water, 62.4 pcf, and the depth to the point under consideration. At 30 ft, this corresponds to 1872 psf. By subtracting the water pressure from the total vertical pressure, the effective vertical pressure is obtained. It is important to note that this effective vertical pressure is the pressure due to the effective weight of the soil particles. To obtain the total pressure, it is necessary to add the water pressure to the effective pressure.

5.3 CAPILLARY STRESSES

The water surface in a brimful glass stands above the lip because of surface tension at the air-water interface. This taut interface can cause water to rise above a free water surface in a small tube and in continuous small soil voids. The magnitude of the surface tension T_s at an air-water interface is nominally 0.075 g/cm.

Figure 5.4 illustrates a hypothetical soil column in a deposit with the free water surface at a-a. A single capillary above a-a is shown shaded to indicate it is filled with water. Other continuous voids will also be filled to some height. The contact between the interface meniscus and soil particles is at some angle α, and the surface tension force acts along that angle. This force acts to pull water above the free surface, a-a. If the effective diameter of the capillary is d, then the total force exerted by the meniscus on the water column is $\pi d T_s \cos \alpha$. Static equilibrium requires that the meniscus support a water column of height z_c, with the result that

$$\pi d T_s \cos \alpha = \frac{\pi z_c d^2 \gamma_w}{4} \tag{5.5}$$

and if $\alpha = 0$, then

$$z_c = \frac{4 T_s}{d \gamma_w} \tag{5.6}$$

Equation 5.6 allows us to estimate the height of capillary rise in soils if we can estimate d, which is related to particle size. We know that d may be approximated by $D_{10}/5$ if D_{10} is taken as the effective size of particles. For instance, for fine sand or silt soils, D_{10} could be estimated as the size of the

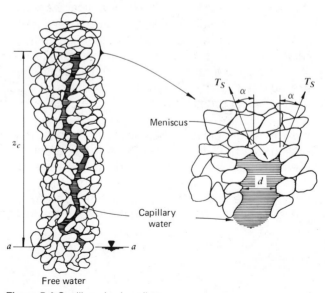

Free water

Figure 5.4 Capillary rise in soils.

opening in the No. 200 sieve, 0.074 mm, or 0.0074 cm. From Equation 5.6

$$z_c = \frac{4 \times 0.075}{(0.0074 \div 5) \times 1} \cong 200 \text{ cm} \tag{5.7}$$

Similar calculations will show that the height of capillary rise is very large in very fine soils and almost negligible in clean, coarse sands and gravels.

Capillary action is important, because it can maintain a high degree of saturation above the water table and produce low soil strength as a result. It also increases effective stress above the free water surface. For instance, force equilibrium at the meniscus in Figure 5.4 requires that

$$\frac{\pi}{4}d^2u + \frac{\pi z_c d^2 \gamma_w}{4} = 0 \tag{5.8}$$

where u is the porewater pressure. From Equation 5.8,

$$u = -z_c \gamma_w \tag{5.9}$$

Thus, in the illustration of Figure 5.3, if the soil had been such that z_c was greater than or equal to 10 ft, the porewater pressure at the ground surface would have been

$$u = -10 \times 62.4 = -624 \text{ psf} \tag{5.10}$$

Since the total stress at the ground surface is zero, the effective stress principle requires that, at the ground surface, the effective stress be 624 psf. With similar calculations, it can be shown that there is a linear increase in effective stress from the free water surface to the height of capillary rise, caused by capillary action. This increment in effective stress is in addition to that due to soil weight. Its importance will be discussed later.

5.4 PERMEABILITY

The permeability of a soil is a measure of the ease with which a particular fluid flows through its voids. Usually, we are concerned with the flow of water through soils. A highly pervious material may be either a headache or a blessing for the contractor. If an excavation below the groundwater level is to be dewatered, a highly pervious soil will require a carefully planned and executed pumping system with comparatively large capacity. Excavation below water in an impervious soil may require no pumping at all. On the other hand, highly pervious material from a borrow source may often be placed and compacted immediately after it is excavated. High permeability permits the excess water to escape almost instantly. Saturated impervious borrow may require days of drying before placement. The extra handling involved is an added expense.

Some ranges of permeability for various soils are given in Table 5.1. These are approximations only but are accurate enough for many purposes. From Table 5.1 it is evident that soils vary markedly in their permeability, with gravels being about 1 million times more pervious than clays. In some instances, it is necessary to determine the permeability coefficient k more accurately than is possible using general descriptions such as shown in Table 5.1. Such determinations may be done in the laboratory, but are often done in the field. In the latter case either borehole permeability tests or pumping tests are conducted. These tests require considerable planning and expense, but offer by far the most reliable information.

TABLE 5.1 TYPICAL SOIL PERMEABILITIES

Soil	Permeability Coefficient, k (cm/sec)	Relative Permeability
Coarse gravel	Exceeds 10^{-1}	High
Sand, clean	10^{-1}–10^{-3}	Medium
Sand, dirty	10^{-3}–10^{-5}	Low
Silt	10^{-5}–10^{-7}	Very low
Clay	Less than 10^{-7}	Impervious

The permeability coefficient k may be considered as the apparent velocity of seepage through a soil under a unit hydraulic gradient. It is numerically equal to the seepage volume per unit of time through a unit area. Analysis of a seepage problem to determine pumping capacity for a dewatering system thus requires not only a representative k for the soil but also a determination of the subsurface flow pattern (area) and associated hydraulic gradients. Such analyses may also provide estimates of uplift pressures under hydraulic structures.

A physical feel for the concept of permeability may be obtained by considering a simple test for its determination. Refer to Figure 5.5, for example. If water is flowing through the device (through the soil) at a constant rate q, we may say that the amount collected Q in a given time period t is

$$Q = qt \tag{5.11}$$

The apparent velocity of flow is

$$v = \frac{q}{A} \tag{5.12}$$

Darcy's law for flow through porous media may be stated as

$$v = ki \tag{5.13}$$

where v is the apparent velocity, k is a constant for the material, the coefficient of permeability, and i is the hydraulic gradient. The latter quantity defined in terms of Figure 5.5 is

$$i = \frac{H}{L} \tag{5.14}$$

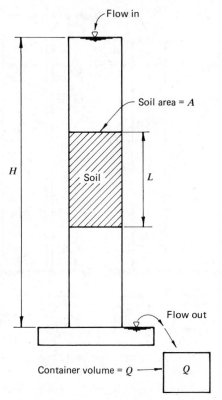

Figure 5.5 A simple constant head permeameter. The inlet and outlet are arranged to maintain constant water surface elevations.

H is the head causing flow over the distance *L*. Combining Equations 5.11 to 5.14, we have

$$Q = vAt$$
$$= kiAt$$
$$= \frac{kH}{L}At$$

which, when solved for *k*, is

$$k = \frac{QL}{HAt} \tag{5.15}$$

Equation 5.15 and the apparatus shown in Figure 5.5 are the basis for laboratory tests to determine the coefficient of permeability for coarse soils. Recognizing the difficulty of sampling such materials and the inherent variability of the deposits from which they come, however, it is considered more reliable to determine these values in the field, when possible.

The layout of a field pumping test is shown in Figure 5.6 for the case of

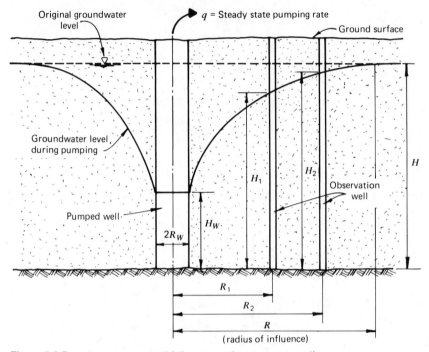

Figure 5.6 Pumping test on a well fully penetrating an open aquifer.

a well penetrating an open soil aquifer. It can be shown that the pumping rate from the well is

$$q = \frac{\pi k (H_2^2 - H_1^2)}{\ln\left(\dfrac{R_2}{R_1}\right)} \tag{5.16}$$

By solving for k, we obtain

$$k = \frac{q}{\pi(H_2^2 - H_1^2)} \ln\left(\frac{R_2}{R_1}\right) \tag{5.17}$$

Thus, by measuring the pumping rate q, and knowing the groundwater depth in two observation wells, we can establish a representative value of k for field conditions. Equation 5.17 may be used to estimate k if no observation wells are available and if it is modified as

$$k = \frac{q}{\pi(H^2 - H_w^2)} \ln\left(\frac{R}{R_w}\right) \tag{5.18}$$

This approach is very approximate, however, in that the radius of influence must be estimated and that the water depth in the well is not a true point on the drawdown curve.

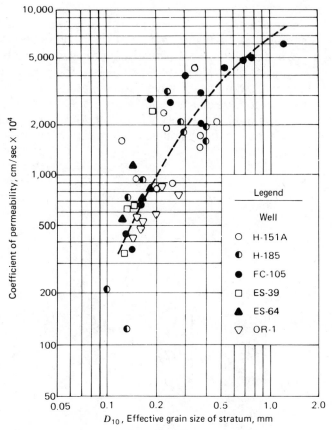

Figure 5.7 Coefficient of permeability of sands determined from pumping tests. (This figure appears through the courtesy of the Corps of Engineers, U.S. Army.)

Some representative values of k from pumping tests are shown in Figure 5.7. It is important to note the magnitude of differences in data point values where the effective sizes of the soil strata are the same.

5.5 COMPRESSIBILITY

Soils subjected to increased effective stress decrease in volume. The stress increase may come about from structure foundations, embankments, or even lowering of the groundwater table. The decrease in volume results in surface settlement. If the soil supports a structure and the settlement is large enough, damage may result. Some soils are very compressible; others are not. If the soil compressibility and loads to be imposed are known, these settlements may be estimated. Good estimates can aid in construction planning where settlements created by construction activities are potentially damaging. This possibility arises where long-term heavy surface loads are applied, both in and around the loaded area and where groundwater lowering is required.

Consider Figure 5.8. The actual soil stratum is represented by column a. It consists of a solid and a liquid phase. This stratum is equivalent to stratum c, wherein the phases have been separated. As the load Δp is applied, the stratum compresses an amount Δh. Since the solid phase is incompressible, the fluid must be expelled from the voids for the settlement Δh to occur. This is indicated by stratum d. If we remember that the void ratio is

$$e = \frac{V_v}{V_s} \tag{5.19}$$

then before the load is applied

$$e_1 = \frac{h_{v1}}{h_s}$$

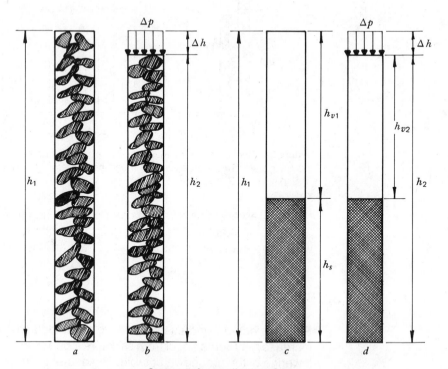

a. Stratum before application of Δp
b. Stratum after application of Δp
c. Idealized stratum a.
d. Idealized stratum b.

Figure 5.8 Idealization of loaded, compressible soil stratum.

if it is assumed that the area of the column is unity. After the settlement is complete

$$e_2 = \frac{h_{v2}}{h_s}$$

and

$$e_1 - e_2 = \frac{h_{v1}}{h_s} - \frac{h_{v2}}{h_s} = \frac{h_{v1} - h_{v2}}{h_s} = \Delta e$$

Now, since the change in height of the stratum is equal to the change in void height,

$$\frac{\Delta h}{h_1} = \frac{\Delta e h_s}{h_s + h_{v1}} = \frac{\Delta e h_s}{h_s + e_1 h_s} = \frac{\Delta e}{1 + e_1}$$

Therefore,

$$\Delta h = \frac{h_1 \Delta e}{1 + e_1} = \frac{h_1(e_1 - e_2)}{(1 + e_1)} \tag{5.20}$$

Equation 5.20 states mathematically that, given the original height and void ratio of a soil stratum and the change in void ratio that might result from a load, we could predict surface settlement. The height in a field situation may be determined by subsurface exploration, and the void ratio may be obtained from tests on samples obtained in the process. The change in void ratio is related to the applied load and is determined for a range of loads in a laboratory consolidation test (ASTM D2435). In the tests, a sample, usually about 1 in. thick and 2½ in. in diameter, is compressed between permeable porous discs. As load is applied, deformation dial readings are taken to record the change in thickness of the sample as water flows from it, through the porous discs. Typical results are shown in Figure 5.9. The slope of the virgin compression curve in Figure 5.9 is

$$C_c = \frac{e_1 - e_2}{\log_{10} p_2' - \log_{10} p_1'} = \frac{e_1 - e_2}{\log_{10} (p_2'/p_1')}$$

By substitution in Equation 5.20,

$$\Delta h = \frac{h_1 C_c}{1 + e_1} \log_{10} \left(\frac{p_2'}{p_1'} \right) \tag{5.21}$$

This relationship is used to predict settlement. The change in effective stress on a soil element, the stratum height, and C_c (the compression index) must be known. The compression index may be considered as a measure of soil compressibility. Most clays in nature are precompressed; that is, they have been loaded in the past to some greater effective stress than they presently carry. For this reason, when they are reloaded, such loading takes

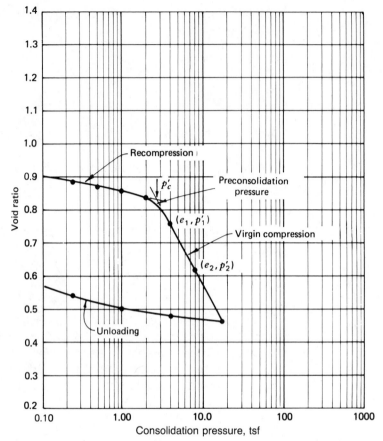

Figure 5.9 Laboratory consolidation test results.

place along the recompression curve shown in Figure 5.9. These clays and also most sands exhibit comparatively low compressibility. Soils that have not been precompressed are much more compressible. Subjected to a change in load, these materials compress along the steeper virgin compression curve. To determine whether a soil has been precompressed, the preconsolidation pressure p'_c (see Figure 5.9) is compared to the vertical effective pressure p'_o at the depths from which the sample was taken. The preconsolidation pressure may be estimated by intersecting tangents to the recompression and virgin compression segments of the pressure-void ratio curve. If p'_c and p'_o are approximately equal, the soil has not been precompressed and is said to be *normally consolidated*. For the case where p'_c exceeds p'_o, the soil has been precompressed and is said to be *overconsolidated*. Identification of potentially compressible soils is discussed in Chapter 4.

It has been noted that compression or consolidation involves expulsion of water from soil voids. It is logical, therefore, that the process should be time dependent. We would not expect it to occur instantaneously. The rate at which consolidation occurs is governed by soil permeability and the efficiency of subsurface drainage provided by the more pervious soil layers. Sands generally are so pervious that any consolidation that occurs does so very rapidly upon application of load. For clays, the consolidation process may take many months or years to complete. Time rate of settlement estimates may be made from information determined in each load increment (each data point on the curve in Figure 5.9) for the consolidation tests mentioned above. Two ways of presenting the same laboratory data for a load increment are shown in Figures 5.10 and 5.11. In both methods the time scale is compressed.

A constant c_v, the coefficient of consolidation, controls the rate of compression of the soil. It is directly proportional to the coefficient of permeability and is defined as

$$c_v = \frac{T_u H^2}{t_u} \tag{5.22}$$

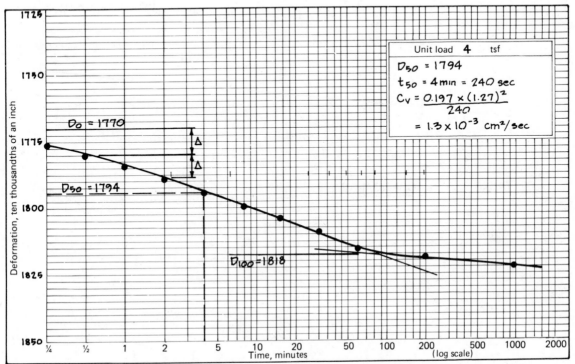

Unit load **4** tsf

$D_{50} = 1794$

$t_{50} = 4\,\text{min} = 240\,\text{sec}$

$c_v = \dfrac{0.197 \times (1.27)^2}{240}$

$= 1.3 \times 10^{-3}\ \text{cm}^2/\text{sec}$

Figure 5.10 Laboratory consolidation test data. Deformation versus logarithm of time.

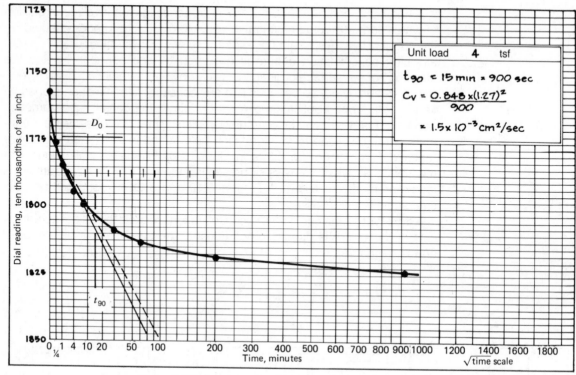

Figure 5.11 Laboratory consolidation test data. Deformation versus square root of time.

where

H = the longest path taken by water seeping from the soils as a result of application of the consolidating pressure increment Δp

t_u = time for a given percentage of consolidation to occur

T_u = a theoretical constant (see Table 5.2)

TABLE 5.2   THEORETICAL TIME FACTORS FOR CONSOLIDATION RATE ANALYSIS

Degree of Consolidation, u (%)	T_u
0	0
10	0.008
20	0.031
30	0.071
40	0.126
50	0.197
60	0.287
70	0.403
80	0.567
90	0.848
100	∞

The value of c_v for a given soil may be determined from the relations for that soil represented by either Figure 5.10 or 5.11 and Equation 5.22. The test data are used to determine the time to a given degree of consolidation.

In Figure 5.10, the semilogarithmic plot of time versus the dial readings during a load increment of the test shows a typical shape. The deformation dial reading, D_{100}, at which primary consolidation is complete (degree of consolidation equals 100 percent) is found by intersecting tangents to the lower segments of the curve. Because the horizontal scale of the plot is logarithmic, there is no zero on the scale at which to determine D_0. It may be located, however, by determining the vertical distance between two points on the initial part of the curve whose times are in the ratio of four to one (Δ as shown), and then extending a horizontal line to the vertical scale, the same distance above the earlier time. As shown in Figure 5.10, the points selected were ½ and 2 minutes. Having established D_0 and D_{100}, the deformation dial reading at 50 percent consolidation is located midway between them, and the corresponding time t_{50} determined. The value of c_v is computed from Equation 5.22, with H equal to the half-sample thickness.

An alternate procedure that employs the same deformation data plotted against the square root of time as shown in Figure 5.11 requires location of the time for 90 percent consolidation. A straight line is drawn through the longest segment of the early part of the curve. A second straight line (the dashed line) is then drawn so that it extends 1.15 times the distance of the first line from the vertical axis. It intersects the curve at t_{90}. Again Equation 5.22 and Table 5.2 are employed to compute c_v. The results should not be greatly different by this method from those obtained by the previous procedure, but exact agreement would not be expected because of the individual judgments that must be made in each.

The values of c_v obtained from laboratory tests are used in time-settlement analyses. There are many limitations on the accuracy of time-settlement analyses, and they are generally considered as estimates only. Predictions of the rate of surface settlement in a given situation may be made, but usually, when it is important to know, field measurements will be made to evaluate these predictions.

5.6   SOIL STRENGTH

The strength of soil materials is a variable and elusive property. Strength characteristics of some materials are represented by simple concepts such as yield point or tensile strength. A given grade of steel is an example. Other materials, such as concrete, are considered in a similar fashion. Concrete strength, though, is very much subject to the quality of its components, their proportions, and the workmanship that goes into its manufacture. The end product is more variable for concrete than it is for steel, even though the strength of both materials is expressed as a simple single number, compressive or tensile strength.

Except for compacted fills, we deal with soils as they are in nature. Because they are not manufactured products, variations in properties are the rule, not the exception. They are natural materials, and in most cases we use what is available, adjusting our designs and working methods to accommodate conditions. Soil strength is not described in a simple single value of tensile strength or compressive strength. There are many kinds of strength tests, some of which apply only to a limited number of soils. In the general case, soil strength may be determined in tests that permit the control of stresses on the test specimen. The triaxial shear test shown in Figure 5.12 is an example. The objective of any soil strength test is to determine the shearing strength of the soil for a given set of conditions. When a soil is loaded to failure in a compression test, as illustrated in Figure 5.13, the results are reduced to a Mohr diagram for interpretation. Stresses at the point of failure are determined. The corresponding Mohr circle is plotted and the strength envelope located. In Figure 5.13 only one Mohr circle is

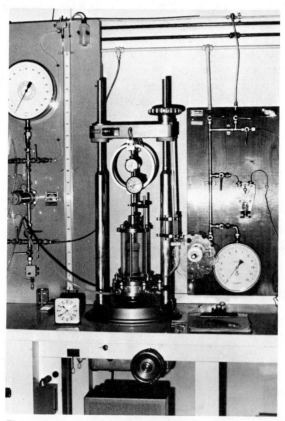

Figure 5.12 Triaxial (cylindrical compression) shear test in progress.

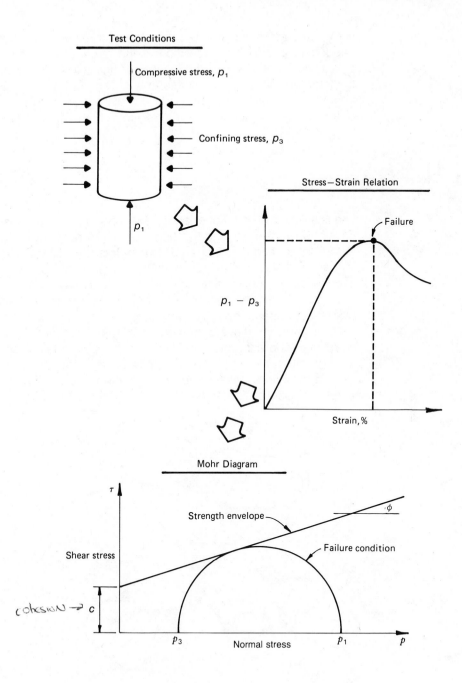

Figure 5.13 Development of soil strength parameters from triaxial test data.

shown, though two or more circles and, therefore, two or more tests, are needed to establish the position of the strength envelope. The strength parameters that describe the location of this line are its slope angle ϕ and the intercept on the shear stress axis c. The angle ϕ is designated as the angle of internal friction, and c is designated as cohesion. The shearing strength s at any pressure is determined from the equation of the envelope:

$$s = c + p \tan \phi \qquad (5.23)$$

5.6.1 Strengths of Sands and Gravels

Sands and gravels are considered to be cohesionless materials. Their strength envelope passes through the origin of the Mohr diagram, and hence the term c in Equation 5.23 is zero. Simply interpreted, the strength equation for sands indicates that they have zero strength if not confined. From our experience we know this is true. Dry or saturated sand on the ground surface has no strength whatever. At some depth, where the material is confined by pressure due to the soil surrounding it, it can support considerable load.

Moist sand (partially saturated) has some cohesive strength or, more correctly, apparent cohesion. This strength is imparted by meniscus forces arising from surface tension in the porewater. Figure 5.14 is a hypothetical section through a moist sand showing the three constituent phases—air, water, and solids. The air-water interface is curved. This curvature and the

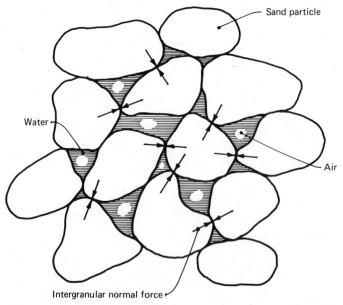

Figure 5.14 Source of strength for moist sands.

surface tension on the interface produce intergranular forces between individual particles. The soil, therefore, has strength since the intergranular normal forces produce frictional resistance. When the soil is dry or saturated the air-water interface is not present, and the apparent cohesion is zero.

The water in a partially saturated mass is in tension. It has been shown that this produces an apparent strength. The water pressure may also be positive. This reduces strength since it reduces intergranular normal forces or effective stresses. When a very loose sand is shaken it tends to densify. If the voids are filled with water, the water must be expelled before densification occurs. When very loose sands are disturbed, the porewater is therefore forced into compression and a positive pressure develops as it attempts to escape. Since the pressure in the fluid increases, the pressure between grains decreases and soil strength is reduced. When the reduction is sufficiently large, the soil has no strength at all and is said to have liquefied. This phenomenon is known as *liquefaction* and it is that condition that exists when a sand or silt is said to be "quick." The same condition can be produced by upward flow of water through a sand bed. This condition can occur in dewatering operations; it is discussed in Chapter 9.

The strength of a cohesionless soil depends on density or unit weight, grain size, shape, mineralogy, fines content, and a number of other factors. Primarily, it is controlled by density. Strength arises from intergranular friction and interlocking controlled by these factors. Variations in strength are indicated by variations in the angle of internal friction. It may vary from 28° to about 45° for a range of conditions from poor to excellent. Generally, cohesionless soils are the most desirable of earth materials. In most cases there is little concern for their strength capabilities.

5.6.2 Strengths of Clays

Clay soils are cohesive in that they possess some strength at zero normal pressure. Consequently, their strength envelope looks like that shown in Figure 5.13. The slope of the strength envelope depends greatly on drainage conditions as the clay is being loaded. For most short-term loading conditions, the usual situation during construction, the slope of the strength envelope is near zero. The strength of the clay is, therefore, expressed by the cohesion term c. For the case in which $\phi = 0$, the pressure difference $p_1 - p_3$ at failure is numerically equal to $2c$ and is most conveniently determined in the unconfined compression test in which p_3 is zero (see Figure 5.13). Unconfined compression test (ASTM D2166) results are the usual measure of clay strength. For a given clay, unconfined strength or cohesive strength will depend strongly on the water content and its relationship to the plastic and liquid limit. Table 5.3 indicates typical ranges of strength corresponding to standard nomenclature for soil consistency. Typical unconfined compression test data are shown in Figure 5.15.

UNCONFINED COMPRESSION TEST DATA

TEST FOR __Northwest Cement__
BY __BJM__
DATE __8/13/71__
SAMPLE IDENTIFICATION __Light Brown Silty Clay (CL)__

PROVING RING NO. AND CONSTANT __542J - 0.34 lb./0.0001 in.__
SPECIMEN WEIGHT __140.89g__ SPECIMEN HEIGHT __2.80 in.__
SPECIMEN DIAMETER
   TOP _____
   CENTER __1.40 in.__
   BOTTOM _____
   AVERAGE __1.40 in.__
AVERAGE SPECIMEN AREA, A_o __1.539 in.^{2}__
AVERAGE LOADING RATE _____
SPECIMEN PROTECTION __Specimen sheared with no moisture protection__

STRAIN DIAL (IN.)	LOAD DIAL (IN.x10^{-4})	LOAD, P (LB.)	AXIAL STRAIN, ϵ	AREA, A $A_o/(1-\epsilon)$ (IN.^{2})	P/A (LB/IN.^{2})	WATER CONTENT DETERMINATION
0	0	0	0	1.539	0	
0.02	30	10.20	.0071	1.550	6.58	Can no. __43__
0.04	86	29.24	.0143	1.561	18.73	Wt. wet __10.58g.__
0.06	127	43.18	.0214	1.572	27.46	Wt. dry __9.33g.__
0.08	152	51.68	.0286	1.584	32.62	Wt. water __1.25g.__
0.10	167	56.78	.0357	1.595	35.59	Tare __3.57g.__
0.12	177	60.18	.0428	1.607	37.44	Wt. solids __5.76g.__
0.14	186	63.24	.0500	1.620	39.03	w% __21.7%__
0.16	191.5	65.11	.0571	1.632	39.89	
0.18	197	66.98	.0643	1.644	40.74	Sketch of Failure
0.20	202	68.68	.0714	1.657	41.44	
0.22	205	69.70	.0786	1.670	41.73	
0.24	211	71.74	.0857	1.683	42.62	
0.26	215	73.10	.0929	1.696	43.10	
0.28	218	74.12	.1000	1.710	43.34	
0.30	220	74.80	.1071	1.723	43.41	
0.32	222	75.48	.1143	1.737	43.45	

REMARKS:
Maximum axial stress = 6390 psf = q_u

Figure 5.15 Unconfined compression test data.

TABLE 5.3   CLAY SOIL STRENGTH AND CONSISTENCY[a]

Consistency	Unconfined Compressive Strength, q_u (tsf)	Cohesive Strength c (tsf)
Very soft	Less than 0.25	Less than 0.12
Soft	0.25–0.50	0.12–0.25
Medium (firm)	0.50–1.0	0.25–0.50
Stiff	1.0–2.0	0.5–1.0
Very stiff	2.0–4.0	1.0–2.0
Hard	Greater than 4.0	Greater than 2.0

Source: K. Terzaghi and R. B. Peck, *Soil Mechanics in Engineering Practice,* Wiley, New York, 1948.
[a] See also Table 7.2.

Soils of stiff to hard consistency seldom present problems when encountered in construction. Very soft to medium clays are very troublesome when they must be supported or excavated and often are unsatisfactory materials for temporary foundations in shoring systems.

5.7   SUMMARY

Soil engineering properties are of concern to contractors only when needed for analysis of their operations. Analysis of problems that involve selection of soil parameters is best done by someone thoroughly familiar with soil behavior and the purpose of the proposed work. Often such an individual is available within the contractor's organization and will usually be a civil engineer. When no person so qualified is available, the contractor should seek outside help. Some classes of problems where engineering properties are important considerations are discussed in later chapters.

The reader should now have obtained an introductory-level understanding of the fundamental engineering properties of soil, permeability, compressibility, and strength. Specifically, the reader should be able to

1   Make static stress calculations, including capillary effects.
2   Define the terms *permeability, compressibility,* and *strength* used in connection with analysis of soil behavior.
3   Describe ways in which soil engineering properties are determined.
4   Compute engineering properties of soils from standard test results.
5   Explain how index properties and engineering properties of soils are related.

REFERENCES

American Society for Testing and Materials, *1974 Book of ASTM Standards,* Philadelphia.

Peck, R. B., Hanson, W. E., and Thornburn, T. H., *Foundation Engineering,* Wiley, New York, 1973.

Terzaghi, K., and Peck, R. B., *Soil Mechanics in Engineering Practice,* Wiley, New York, 1967.

U.S. Army Corps of Engineers, *Investigation of Underseepage and Its Control, Lower Mississippi River Levees,* Waterways Experiment Station TM 3-424, Vicksburg, Miss., 1956.

PROBLEMS

1 The sequence of soils at a site is shown in Figure 5.16. Calculate the total pressure, water pressure, and effective pressure on horizontal

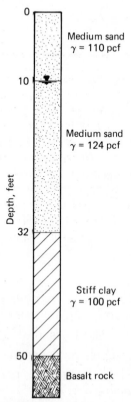

Figure 5.16 Soil profile for Problem 1

planes at each change of soil type, and at the groundwater table. Show your results on a sketch, indicating the variations of each of these pressures with depth.

2 Solve Problem 1 if the sands overlying the clays were silts with the same densities. State any assumptions you make.

3 Estimate the height of capillary rise in a clay soil.

4 A series of three experiments was run on a single sample using apparatus like that shown in Figure 5.2. Q was 60 cm³, L was 12 cm, and A was 25 cm². Paired values of H and t are in the following table:

H (cm)	t (sec)
20	140
40	73
60	50

Demonstrate the validity of Darcy's law (Equation 5.3) in this instance. Estimate the coefficient of permeability of the sample.

5 A 6-in.-diameter well that fully penetrates a 40-ft-thick open aquifer is pumped at a rate of 100 gpm. The drawdown in the well is 22 ft when a steady water level is established. The drawdown in a well 152 ft distant is 12 ft. Estimate the coefficient of permeability of the soil deposit.

6 Discuss the validity of the answer obtained in Problem 5. Suggest how you would improve the test data obtained to produce a better value of k. Why?

7 Data from a one-dimensional consolidation test are presented in the following table. Plot the void ratio-pressure relationship and estimate the compression index for both the virgin compression curve and the recompression curve.

Consolidating Pressure (tsf)	Void Ratio
1/8	1.87
1/4	1.86
1/2	1.84
1	1.81
2	1.76
4	1.66
8	1.51
16	1.32
4	1.36
1	1.44
1/4	1.49

8 Data from a single load increment during a consolidation test on a ¾-in.-thick sample with drainage on both ends follow:

Elapsed Time (minutes)	¼	1	4	8	16	30	60	120	240	480	1200
Deformation Dial (0.0001 in.)	1600	1582	1544	1510	1465	1410	1330	1269	1232	1210	1200

Determine the coefficient of consolidation using both the logarithm of time and square root of time methods.

9 Data from unconfined compression tests on a sample (one test on the undisturbed sample, one on the same sample remolded) are shown in the following table. Plot the axial stress versus strain curves. Estimate the unconfined compression strength and soil cohesion in each case, in tons per square foot. What is the effect of remolding? Why?

Axial Strain (%)	Undisturbed Axial Stress (psi)	Remolded Axial Stress (psi)
0	0	0
2	6.1	1.4
4	6.8	2.5
6	6.9	3.2
8	6.9	3.7
10	6.8	4.2
12	6.7	4.5
14	6.6	4.7
16	6.5	4.8
18	—	5.0
20	—	5.1

10 The major principal stress at failure for a triaxial compression test on dense sand was 5.15 tsf. The minor principal stress, or lateral pressure, was 1.45 tsf. Plot the Mohr circle at failure and, assuming the sand to be cohesionless, determine the angle of internal friction.

11 Calculate the unconfined compressive strength of a moist, fine sand. Carefully state any assumptions you might make. Illustrate your calculations with sketches if appropriate.

PART TWO

EARTHWORK IN THE CONSTRUCTION CONTRACT

CHAPTER 6

THE CONTRACT AND CONTRACT DOCUMENTS

In any construction work under contract, the agreement between the owner and the contractor regarding the work to be done is contained in the contract documents. These documents also include certain other information that establishes the basis for the agreement and aids in the administration of the contract. The documents are prepared by the owner's representative, who is usually an engineer or architect, unless the owner serves in this capacity. The responsibilities of all parties are explicitly put forth. Since any contract is practically limited in format and scope only by the imagination of the parties involved, no discussion of these matters is ever all-encompassing. The sections that follow do, however, contain information related to standard features of most contract documents. The intent of this chapter is to call attention to those features specifically treating provisions affected by soil conditions. Figure 6.1 shows that the number of such items may be large. The chapter includes

1   A discussion of the relationships and responsibilities among parties to contracts.

2   Enumeration of contents of typical construction contract documents.

3   An example special specification for execution of a particular portion of the earthwork in a project.

4   A section on settling disputes.

5   Coverage of the "changed condition" situation where soil conditions encountered in construction are different from those anticipated.

Figure 6.1 A major construction project involving several operations related to soil behavior. These include excavation and backfilling, dewatering, excavation supports, shallow foundation and pile foundation construction, and soil stabilization.

6.1 PARTIES TO THE CONTRACT

Administration of a construction contract principally involves three parties. They are the owner, the contractor, and the engineer.* It is the owner who needs the work done. The contractor executes the work according to plans prepared by the engineer. The engineer, having prepared the plans, usually is best suited to see that the owner's interests are served during construction and, therefore, is present on the job during construction as the owner's representative. The contract, however, is between the owner and the contractor.

The relationships among contracting parties are shown in Figure 6.2. The three parties' responsibilities are divided according to their benefits. The owner needs the work done and, therefore, must pay the contractor to do it and pay the engineer to assist with planning and construction. The owner's staff will generally oversee the progress of the work of both the contractor and the engineer. Some owners become closely involved in the work while

---

* Occasionally, we have used masculine pronouns to refer to the engineer, contractor, or owner. We have done so for convenience and succinctness.

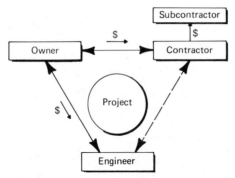

Figure 6.2 Contractual relationships in private contracts.

others prefer the engineer to administer their interests in all phases of the project.

The contractor is obligated to the owner, but deals mostly through the engineer. He employs subcontractors to assist with specialty work and is responsible for their work as if they were part of his own forces. Subcontractors, in principle, do not deal directly with the owner or the engineer. Their contract, if they have one, is directly with the contractor.

The engineer is not a party to the construction contract. He has no financial stake in its outcome and, thus, is in the best position to administer the contract in a situation where the owner wants to minimize expense and the contractor wants to maximize profit. The principal disadvantage of the relationships shown in Figure 6.2 lies with the fact that the contractor is sometimes asked to deal with two parties rather than one. The owner and engineer generally should speak with one voice. Unfortunately, this is not always the case.

Figure 6.3 illustrates the contractual relationships in which the engineer and owner are one and the same. Such situations are common when the owner maintains a large engineering staff. Government projects are a good example. The arrangement offers the advantage of being streamlined ad-

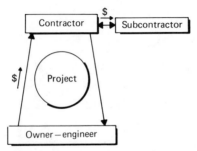

Figure 6.3 Contractual relationships with an owner-engineer.

ministratively. But, in this instance, the engineer's interests are more closely tied to those of the owner than in the previous case. In the three-party relationship the contractor can appeal to the owner in the event of a dispute with the engineer. In the two-party relationship there is no one else to talk to.

Construction work of some types is being increasingly handled through the construction management process. The owner's representative in this situation is the construction manager. Frequently, the construction manager may be the contractor, although that is not a requirement. He may also be the engineer, in which case the relationships are essentially as shown in Figure 6.3. Whoever serves as construction manager must be knowledgeable enough of both the design and construction functions to offer the owner the advantage of this method of doing business. If we presume an independent construction manager, the relationships among the parties are as shown in Figure 6.4. Proponents of this method will argue that it speeds construction because the manager may initiate some phases of work before completion of design, since the contractor will have been selected. Because the construction expertise of the contractor is available, his knowledge may also have an economizing effect on design.

6.2 DOCUMENT CONTENTS

In arranging for the selection of a contractor under the usual method (see Figure 6.2), the owner for whom the proposed work is to be done will advertise his intention to accept bids. This advertisement will contain a very general description of the work and direct prospective bidders to obtain

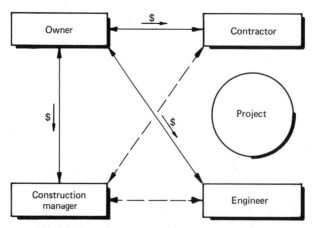

Figure 6.4 Contractual relationships established by the construction management method.

further information regarding the project from the designated source, usually the engineer. Prospective bidders will be provided with a set of contract documents upon application. These documents specify the work to be done and administrative procedures and arrangements among the parties involved, and include also the contract agreement and bidding forms and requirements. The agreement and work will be governed entirely by this document. The usual essential features of the documents are included in the following sections:

Advertisement for bids
Proposal (bid form)
Contract
Information for bidders
General conditions
Special specifications
Drawings

The work to be accomplished is detailed in the drawings and further clarified by the special specifications. The quantities and prices agreed on are listed in the proposal. The general conditions outline the procedures by means of which the project will be administered and define the duties and responsibilities of the parties concerned. Also contained are bonding and insurance requirements for the contractor. In many cases, the special specifications and drawings will include information concerning site subsurface conditions that may be available. Often, however, this information will not be included in the contract documents. It will be referenced therein as available in the engineer's office for inspection. Such arrangements are intended to preclude the possibility that subsurface conditions anticipated become part of the contract by inference.

6.3 BID PREPARATION

Having read the advertisement for bids and determining that he may be interested in bidding, the contractor obtains a set of documents from the designated source. After examining the work to be done, he may prepare and submit the proposal (bid) contained therein. In his proposal, the contractor lists the prices for which he will perform each item of required work. By submitting a proposal, the contractor certifies that he understands the documents, and that if selected as the successful bidder he will do the work as proposed. Excerpts regarding site conditions from a standard proposal follow.

The Bidder further declares that he has carefully examined the Contract Documents for the construction of the project; that he has personally inspected the site; that he has satisfied himself as to the quantities involved, including materials and equipment, and conditions of work involved, including the fact that the description of the quantities of work and materials, as included herein, is brief and is intended only to indicate the general nature of the work and to identify the said quantities with the detailed requirements of the Contract Documents; and that this Proposal is made according to the provisions and under the terms of the Contract Documents, which Documents are hereby made a part of this Proposal.

The Bidder further agrees that he has exercised his own judgment regarding the interpretation of subsurface information and has utilized all data that he believes pertinent from the Engineer, Owner, and other sources in arriving at his conclusions.

By submitting the proposal containing these statements, the contractor has in effect said, among other things, that he understands site conditions and their effects on his work well enough to submit a valid bid. He then proceeds to itemize his costs. A section of a proposal for a unit price contract on excavation and embankment is given in Figure 6.5. The proposal of the successful bidder is incorporated in the contract, along with the contract documents. The contract provisions recognize that there may be circumstances under which the proposed costs may change. These circumstances are noted in the contract itself and are fully defined elsewhere in the documents. Such provisions are a part of all construction contracts, as in the following excerpt, and are not restricted to earthwork alone.

In consideration of the faithful performance of the work herein embraced, as set forth in these Contract Documents, and in accordance with the direction of the Engineer and to his satisfaction to the extent provided in the Contract Documents, the Owner agrees to pay to the Contractor the amount bid as adjusted in accordance with the Proposal as determined by the Contract Documents, or as otherwise herein provided, and based on the said Proposal made by the Contractor, and to make such payments in the manner and at the times provided in the Contract Documents.

Quantities also may vary. This fact is called to the attention of the contractor in the information for bidders section. In some cases, contracts are written so that a quantity change exceeding a specified percentage of the total item is justification for negotiation of new unit prices. The information for bidders section usually covers these provisions.

ITEM	QUANTITY	UNIT OF MEASURE	UNIT PRICE FIGURES	UNIT PRICE IN WRITING	TOTAL AMOUNT QUANTITY x UNIT PRICE
2A. Earthwork					
Unclassified Excavation					
	79,900	cu.yds.	$ 0.96	$ Ninety six cents	$ 76,704
Removal of Waste Material					
	5,800	cu.yds.	$ 2.00	$ Two dollars	$ 11,600
Classified Embankment from Unclassified Excavation					
Zone 1	33,000	cu.yds.	$ 0.86	$ Eighty Six Cents	$ 28,380
Zone 3	16,500	cu.yds.	$ 0.92	$ Ninety two cents	$ 15,180
Classified Embankment Imported					
Zone 1	43,000	cu.yds.	$ 1.20	$ One dollar and twenty cents	$ 51,600
Zone 2	37,500	cu.yds.	$ 1.20	$ One dollar and twenty cents	$ 45,000
Zone 3	5,500	cu.yds.	$ 1.20	$ One dollar and twenty cents	$ 6,600
Zone 4	37,250	cu.yds.	$ 2.00	$ Two dollars	$ 74,500
Zone 5	3,000	cu.yds.	$ 1.20	$ One dollar and twenty cents	$ 3,600
				TOTAL	$ 315,164

Figure 6.5 A unit price proposal for excavation and embankment.

When the Proposal for the work is to be submitted on a unit price basis, unit price proposals will be accepted on all items of work set forth in the Proposal, except those designated to be paid for as a lump sum. The estimate of quantities of work to be done is tabulated in the Proposal and, although stated with as much accuracy as possible, is approximate only and is assumed solely for the basis of calculation upon which the award of Contract shall be made. Payment to the Contractor will be made on the measurement of the work actually performed by the Contractor as specified in the Contract Documents. The Owner reserves the right to increase or diminish the amount of any class of work as may be deemed necessary, unless otherwise specified in the Special Provisions.

6.4 CONTRACT EXECUTION

When the successful bidder has been determined and the contract executed, the work is undertaken as called for by the documents. The owner will be represented on the project by a resident engineer whose principal function is to ensure that the work complies with specifications. Each general class of work shown on the drawings is covered by a special specification section. Each special specification section is in turn divided into four categories: scope, materials, workmanship, and payment. Their content and purpose in the earthwork section are illustrated by the following paragraphs, excerpted from a contract.

---

2. EARTHWORK

A. Scope.

This section covers the work necessary for the earthwork, complete.

B. Materials

Unclassified Excavation.   Excavation is unclassified. Complete all excavation regardless of the type, character, nature, or condition of the materials encountered. The Contractor shall make his own estimate of the kind and extent of the various materials to be excavated in order to accomplish the work.

Classified Embankment.   Material from ***Unclassified Excavation*** shall be used wherever possible. Additional materials required shall be imported to the project site.

Embankment Zone 1.   Clean granular material uniformly graded from coarse to fine; less than 12 percent finer than No. 200 U.S. standard sieve; maximum size, 3 in. Submit samples to Engineer for approval.

Embankment Zone 2.   Clean granular material uniformly graded from coarse to fine; less than 5 percent finer than No. 200 U.S. standard sieve; less than 10 percent finer than No. 4 U.S. standard sieve; maximum size, 8 in. Submit samples to Engineer for approval.

Embankment Zone 3.   Granular material; less than 30 percent finer than No. 200 U.S. standard sieve. Submit samples to the Engineer for approval.

Embankment Zone 4.   As specified for *Embankment Zone 2.*

Embankment Zone 5.   Sandy loam free from clay, roots, organic matter, and gravel; suitable for topsoil. Submit samples to Engineer for approval.

Waste Material.   Material from *Unclassified Excavation* found unsatisfactory for use in *Classified Embankment* shall be designated *Waste Material.*

C. Workmanship

General.   The Contractor shall at all times conduct his work so as to ensure the least possible obstruction to water-borne traffic. The convenience of water-borne traffic and the protection of persons and property are of prime importance and shall be provided by the Contractor in an adequate and satisfactory manner. If the Contractor blocks the channel, he shall remove such obstructions to a width sufficient to allow safe passage, as the water-borne traffic approaches.

The limits of the area to be dredged at the site are shown on the Drawings. The Contractor shall assume full responsibility for the alignment, elevations, and dimensions of the area to be dredged.

Dredging Equipment.   Dredging equipment shall be of adequate capacity to perform the work within the designated time. Dredge may be a suction type, hydraulic type, or clam shell bucket type, or dragline at the Contractor's option.

Tolerance.   Dredge to elevations and cross sections indicated on Drawings. Do not dredge to depths greater than needed or shown. A tolerance of one foot, plus or minus, from the elevations shown will be allowed. Dredging shall proceed from the top to the bottom of the slope in such a manner that the inclination of the slope does not exceed that indicated. Work laterally on slope to prevent local oversteepening.

Construction of Embankment Area with Dredged Material Conforming to Material Specification.   The dredged material shall be deposited over the embankment area within the limits indicated on the Plans. The embankment shall be constructed to the lines, grades, and cross sections shown on the Plans or established by the Engineer. Failure to conform to the established lines and grades will not be tolerated. Construct all dikes, ditches, culverts, and other related and incidental work necessary to place and drain the embankment as required for a satisfactory job.

If a washout of dredge deposited material should occur, or for any other reason material is deposited outside of the designated area, except those silts and clays that remain in colloidal suspension, this material shall be removed and placed in the designated area at no additional cost to the Owner.

Dredged embankment that is lost prior to completion and acceptance of the work shall be replaced by the Contractor at no additional cost to the Owner. All dredged areas and embankments shall conform to the requirements of the Drawings and Specifications within the tolerances specified.

Grading.   Grade the fill area to a smooth surface. Side slopes shall be within 0.5 ft of true grade.

Embankment Zone 1.   Place in lifts with maximum compacted thickness of 1 ft. Compact to at least 70 percent relative density according to ASTM D2049.

Embankment Zone 2.   Place in uniform lifts. No compaction shall be required. Maximum lift thickness shall be 2 ft.

Embankment Zone 3.   Place in uniform lifts. No compaction shall be required. Maximum lift thickness shall be 2 ft.

Embankment Zone 4.   Place in accordance with Plans. No compaction shall be required. Place from bottom of slope to water surface.

Embankment Zone 5.   Place in uniform lifts. No compaction required. Maximum lift thickness shall be 1 ft.

Waste Material.   Remove from project site.

Rate of Excavation and Filling.   Excavation from the river bank and placement of embankment materials shall take place at rates that will not impair the stability of the river bank slope. It shall be the responsibility of the Engineer to detect the development of conditions that could lead to instability of the river slope which arise from excessive rates of excavation or filling. The Engineer shall notify the Contractor to stop work or modify procedures as necessary until such conditions as may develop, subside, or are otherwise corrected. This section shall not be interpreted as relieving the Contractor of the responsibility for the consequences of failure to carry out excavation of the slope as specified under *Tolerances* above.

Compaction Control.   The Contractor shall be primarily responsible to assure that the compaction specifications for *Embankment Zone 1* material have been met. Tests and measurements shall be made by the Engineer to verify the compaction obtained.

D. Payment

Unclassified Excavation.   Payment for excavation will be based on the unit price per cubic yard stated in the Contractor's Proposal and the number of cubic yards satisfactorily excavated within the authorized limits. No payment will be made for material dredged below the established grade line. Measurement shall be by cross section taken prior to excavation by the Engineer.

Waste Material.   Additional payment shall be allowed for removal from the site of *Unclassified Excavation* unsuitable for use in the embankment sections. The additional payment shall be at the unit price stated in the Contractor's Proposal. Measurement shall be made in the haul units at the time the material is removed from the site.

Classified Embankment. Payment for this item shall be at the unit prices stated in the Contractor's Proposal. Payment shall be based on the number of cubic yards satisfactorily placed within the authorized limits. Measurement shall be by cross section taken of the completed work by the Engineer.

Standby Time. If a work stoppage or curtailment is necessary under the provisions of ***Instrumentation*** or ***Rate of Excavation and Filling*** above, payment shall be made to compensate the Contractor for equipment standby time. Payment shall be at the rate of 20 percent of the rate for individual equipment items stated in the current issue of Rental Rates for Equipment without Operators, published by the Oregon State Highway Department. Payment shall be made only for those items of equipment in use at the time of stoppage or curtailment and which remain on the project site and are unused during the period of delay.

In this specification, the requirements for dredge excavation and subsequent site filling are set forth. When reading this section, the contractor learns first that this section covers all specifications for the earthwork noted in his proposal. The material in the excavation is then identified. The embankment material requirements are given. The contractor is given certain restrictions on how he may proceed (or not proceed) with the work. In this regard, it should be noted that well-written specifications will not be prepared so as to assume for the owner the responsibility for the proper conduct of the contractor's work. Finally, a method of payment is specified. Thus, by studying the special specifications, the contractor may determine the technical requirements and payment provisions for each item in the proposal.

6.4.1 Disputes

A contract completed without some dispute is rare. In contracts involving soil materials, disputes frequently arise over the effect of site conditions on the progress of the work. In preparing a bid the contractor has had to consider site conditions. What consideration he has actually given may range from almost none, to very detailed, careful study. In any case when bad site conditions substantially affect the contractor's profit picture, a dispute over what site conditions were represented to be will almost certainly arise.

For whatever cause they arise, disputes are most frequently settled by mutual agreement. Long arguments, voluminous paperwork, and a certain amount of bluffing usually lead one of the parties to the conclusion that his position is untenable, or that a compromise may be worked out. Settling a dispute in this manner frequently produces a change order (formal change in the contract) agreed to by both parties. For instance, when rock excavation is required, but had not been anticipated, and when it can be shown that a reasonable contractor should not have expected it, a method of payment for the work will have to be agreed on, and a change order may be issued. The settling of disputes by agreement in this manner offers the advantage of timeliness, in that it is done while the work is in progress.

Disputes not settled by mutual agreement may be settled by mediation, arbitration or litigation. In each instance disinterested third parties decide the issues. Litigation is especially costly and apt to take years to conclude. It is, therefore, appropriate only when all other methods have failed. Recently, mediation and arbitration have become popular methods of resolving construction disputes. They offer certain advantages over litigation, in addition to progressing more rapidly and being generally less costly.

Arbitration is a procedure in which the parties in the dispute select a third party to whom arguments will be presented, and who will decide the issue. Proceedings are conducted similarly to courtroom litigation, except that there is no jury, and the rules of evidence are less formal. The arbitrator often questions witnesses, along with counsel for both sides. The arbitrator's decision is usually binding.

Mediation is an even less formal procedure in which the disputing parties agree on a third party who will examine opposing views and try to work out a fair solution or compromise. The mediator is essentially a go-between who develops a position that he hopes both sides can live with. This procedure may fail, in which case the dispute may end up in arbitration or in court.

6.4.2   Changed Conditions

It is near-universal practice to include in contract documents statements that require the contractor to indicate that he has carefully examined all subsurface information available for the project site, and that he knows how these conditions may affect his work. He further must acknowledge that he can do the work, in light of these conditions, for the bid price. An example is given below.

Site Investigation and Representation.   The Contractor acknowledges that he has satisfied himself as to the nature and location of the work, the general and local conditions, particularly those bearing upon availability of transportation, disposal, handling and storage of materials, availability of labor, water, electric power, roads, and uncertainties of weather, river stages, or similar physical conditions at the site, the conformation and conditions of the ground, the character of equipment and facilities needed preliminary to and during the prosecution of the work and all other matters that can in any way affect the work or the cost thereof under this Contract. The Contractor further acknowledges that he has satisfied himself as to the character, quality, and quantity of surface and subsurface materials to be encountered from inspecting the site, all exploratory work done by the Owner, as well as from information presented by the Drawings and Specifications made a part of this Contract. Any failure by the Contractor to acquaint himself with all the available information will not relieve him from responsibility for estimating properly the difficulty or cost of successfully performing the work. The Contractor warrants that as a result of his examination and investigation of all the aforesaid data that he can perform the work in a good and workmanlike manner and to the satisfaction of the Owner.

It is usually further noted that the available subsurface information may not be sufficient to fully describe conditions as they exist.

Subsurface and Site Information.   Test holes have been excavated to indicate subsurface materials at particular locations. This information is shown on the Drawings. Investigations conducted by the Engineer of subsurface conditions are for the purpose of study and design, and neither the Owner nor the Engineer assumes any responsibility whatever in respect to the sufficiency or accuracy of the borings thus made, or of the log of test borings, or of other investigations, or of the interpretations made thereof, and there is no warranty or guarantee, either expressed or implied, that the conditions indicated by such investigations are representative of those existing throughout such area, or any part thereof, or that unforeseen developments may not occur.

By virtue of these paragraphs the contractor is required to certify that he has fully informed himself using available information and that he is also aware that this information may be inadequate to describe site subsurface conditions. He is also asked to certify that, in spite of this possibly inadequate knowledge, he can do the work for the price bid. Such language is used in some cases in an attempt to burden the contractor with the risks inherent in the site. Usually, however, its purpose is to protect the owner against frivolous claims.

When difficult ground conditions are encountered during a project they may be fully predictable from information made available to the contractor and may be readily dealt with by the methods planned. On the other hand, they may be a surprise to everyone concerned and the planned construction methods may be inadequate. The former case is what should transpire if everyone has done his job well; the engineer in representing the site, and the contractor in interpreting its effects on his work. The latter case may develop in spite of everyone's best efforts, simply because natural ground conditions are not totally predictable. In this case the owner is usually obligated to bear the additional costs involved in developing his site, in spite of contractual disclaimers of responsibilities like those described earlier. This is a true change in conditions. Competent people did not expect or anticipate the actual situation, but if they had, the owner would have had to bear the costs involved.

There is also the possibility that a contractor losing money because of difficult conditions may claim that these conditions are different than he expected, and that he is therefore entitled to extra compensation. For his claim to stand, he must show that conditions were substantially different than he reasonably could have expected, considering the information available and his knowledge and capability as a contractor.

In the event of a potential changed condition claim it is imperative that the facts of the matter be fully documented in a timely fashion. To this end, the best available help should be sought at an early stage. Substantial amounts of money are often involved, and the positions of either side of the dispute will be strengthened by immediate and thorough attention to the problem.

6.5   SUMMARY

Items in earthwork specifications may be many and varied. This chapter has illustrated such specifications by example. Earthwork in construction contracts is often the subject of dispute and controversy, partly because the materials involved, soil and rock, have properties that are not always correctly anticipated. These disputes are most difficult to resolve when the parties involved are unwilling to assume their proper responsibilities for describing site conditions and assessing their effects on construction operations. Having studied the chapter, the reader should

1   Be able to describe relationships among contracting parties and their agents and the responsibilities of each.

2   Know the contents of the usual construction contract.

3   Understand what constitutes a complete special specification.

4   Know how contract disputes are settled.

5   Understand what "changed conditions" means and the circumstances under which a contract change order may result from them.

REFERENCES

Abbett, R. W., *Engineering Contracts and Specifications,* Wiley, New York, 1967.

Canfield, D. T., and Bowman, J. H., *Business, Legal and Ethical Phases of Engineering,* McGraw-Hill, New York, 1954.

Sadler, W. C., *Legal Aspects of Construction,* McGraw-Hill, New York, 1959.

Sowers, G. F., "Changed Soil and Rock Conditions in Construction," *Journal, Construction Division, American Society of Civil Engineers,* **97,** No. CO2, 1971.

CHAPTER 7

INTERPRETATION OF SOILS REPORTS

Soils explorations are usually made in advance of construction to assist an engineer in preparing a design. In most cases, the reports of these explorations will substantially assist a contractor in preparing a bid. The usual subsurface exploration and soils investigation is done, however, without the specific needs of the contractor in mind. When the private design engineer has completed his exploration and investigation, he prepares a report that he submits to his client who will be the owner in the eventual construction contract. On government projects where engineer and owner are one, such work is often handled with internal memoranda. These reports and memoranda contain a great deal of information that will be useful to the contractor in assembling information with which to prepare a bid. The reports and memoranda and all other information that relate to them will usually be made available to the contractor in the period between bid advertisements and bid opening. Selected information may be included in the contract documents. Most of it is referenced therein. The wise contractor will obtain this information, study it, interpret it, and consider it in the preparation of his bid. This chapter is intended to be a guide to use of subsurface exploration reports. The objectives of the chapter are to

1 Provide a clear definition of the purposes of the usual subsurface exploration.

2 Illustrate the various methods of subsurface exploration available, the advantages and disadvantages of each, and the form and utility of presentation of results from each.

3 Describe the form and format of the usual soils engineering report.

4 Suggest sources of available subsurface information.

7.1   FIELD EXPLORATIONS

Field explorations may be considered to have three fundamental objectives. The first of these is to locate and define the boundaries of the various soil and rock strata that underlie the site of the proposed construction. The second is to locate the groundwater table. Finally, it is necessary to determine the engineering properties of the subsurface materials. This may be done by field testing during the exploration or by testing of samples obtained during the exploration and returned to the testing laboratory. The third objective of the exploration then would be to perform field tests or to secure the samples required for laboratory testing so that the engineering properties of the various strata might be determined. The methods used in accomplishing these objectives will vary according to the manner in which the results of the exploration will be used.

7.1.1   Direct Exploration

The most reliable information may be obtained by directly observing the subsurface materials at a specific site. Direct observation requires excavation of test pits, test trenches, or test holes that are large enough to permit access for visual inspection. In the usual case, direct exploration is limited to fairly shallow depths. Where the groundwater table is not present excavation may be readily accomplished to depths up to 20 ft using commonly available construction equipment. Deeper excavation for direct observation requires mobilization of large-diameter drilling equipment or hand mining. These techniques will be used only rarely. When the groundwater table is encountered near the surface, direct exploration techniques are usually discarded in favor of some other method.

The principal advantage of direct exploration is that it permits direct observation of subsurface materials. Soil and rock materials may be observed in their natural state. Minor details of subsurface profiles that may have great significance in engineering and building on a site may be readily seen. Undisturbed samples of subsurface materials may be obtained by removing them at the level desired. Large disturbed samples are also easily obtained. The principal disadvantage of direct exploration is that it is usually restricted in depth. If the depth of the exploration is not restricted by the available equipment or groundwater conditions, then the cost becomes an important factor in deciding whether to use this method.

7.1.2   Semidirect Exploration

To overcome the restrictions inherent with direct exploration, semidirect techniques have become highly developed. These methods are in fact the most widely used techniques for engineering subsurface investigations. A semidirect exploration involves drilling and sampling. A boring is made using one of many methods to gain access to desired locations at depth. Samples are obtained at these locations with techniques that are determined by the purposes for which the samples will be used and the conditions under which they are taken. A typical auger drilling setup is shown in Figure 7.1. The power-driven auger may have either a solid or a hollow stem. Solid-stem augers are used typically in dry regions where the groundwater table will not be encountered. This is because the auger must be removed from the hole prior to attempting to obtain a sample at any elevation. In the event that the groundwater table is high and the subsurface materials are loose, the removal of the auger usually results in the collapse of the hole. Hollow-stem augers were developed to permit sampling through the stem of the auger without removing the auger from the boring. The drilling and sampling operation was thus improved since the auger could be left in place. The operation was faster and the stem could function as a casing, always stabilizing the sides of the borehole. Rotary drilling equipment such as that

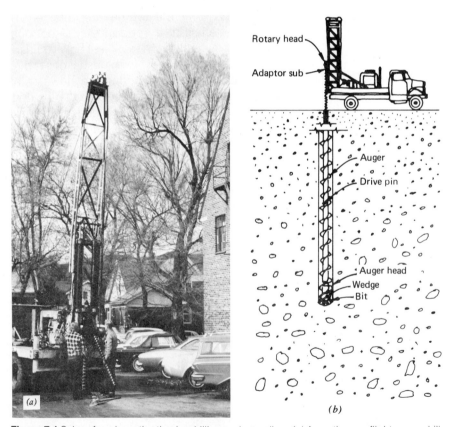

Figure 7.1 Subsurface investigation by drilling and sampling. *(a)* A continuous flight auger drill. *(b)* Schematic drilling setup (Longyear Company, Minneapolis, Minn.)

shown in Figure 7.2 is also commonly used. In this method a clay slurry or driller's mud is circulated down the center of the drill rod and out the end of the roller bit. This slurry picks up the cuttings from the roller bit as it rotates and circulates along the outside of the drill stem returning to the ground surface. The cuttings are deposited in a mud pit and the slurry is returned by way of the pump through the drill stem. A continuous fluid system is thus established, which removes the material cut as the boring advances and stabilizes the sides of the borehole.

When the boring reaches the elevation at which a sample is to be taken, one of several sampling methods may be used. Split barrel sampling or standard penetration testing (ASTM D1586) involves driving a standard sampler with a standard amount of driving energy. A test in progress is shown in Figure 7.3. The resistance to penetration of the sampler is an index of the consistency or relative density of the ground. This standard split barrel

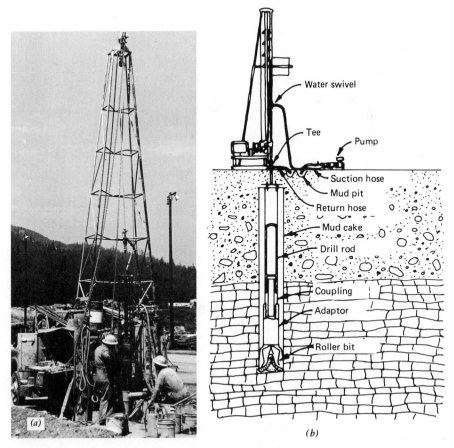

Figure 7.2 Rotary hydraulic drilling equipment. *(a)* Truck-mounted rotary hydraulic drill. *(b)* Schematic rotary hydraulic drill. (Longyear Company, Minneapolis, Minn.)

TABLE 7.1   RELATIVE DENSITY OF COHESIONLESS SOILS

Relative Density	N — Standard Penetration Resistance
Very loose	0–4
Loose	4–10
Medium	10–30
Dense	30–50
Very dense	Over 50

Source: K. Terzaghi and R. B. Peck, *Soil Mechanics in Engineering Practice,* Wiley, New York, 1948.

sampler is 2 in. in outside diameter and about 30 in. long. Either end of the sampler may be removed and the center section, which is split, opened to reveal any sample that might be retained. The sample will be 1⅝ in. in diameter and heavily disturbed. It is suitable usually only for classification and identification purposes. The driving weight weighs 140 lb and is dropped a distance of 30 in. where it strikes the collar on the sampling rod. The number of blows struck in this manner that it takes to drive the sampler three consecutive 6-in. increments is recorded. The number of blows taken to drive the final two 6-in. increments is totaled and designated as N, the standard penetration resistance. Standard penetration resistance has been correlated in a rough way with the properties of cohesionless and cohesive soils, as indicated in Tables 7.1 and 7.2. These values should be considered as approximations only. Other drive sampling methods are used, involving heavier weights, different drops, and larger samplers, and their results have been correlated, in a rough way, with the results of standard penetration tests. All penetration tests, however, do not produce directly comparable results.

In contrast to split barrel sampling, thin-walled tube sampling (ASTM D1687) is accomplished by shoving rather than driving a sampling tube into the bottom of a boring. These tubes will typically be 2 to 5 in. in diameter and up to 4 ft in length. The purpose of thin-wall tube sampling is to obtain specimens for laboratory testing that are undisturbed. No usable information

TABLE 7.2   CONSISTENCY OF COHESIVE SOILS[a]

Consistency	N — Standard Penetration Resistance
Very soft	Less than 2
Soft	2–4
Medium (firm)	4–8
Stiff	8–15
Very stiff	15–30
Hard	Over 30

Source: K. Terzaghi and R. B. Peck, *Soil Mechanics in Engineering Practice,* Wiley, New York, 1948.
[a] See also Table 5.3.

Figure 7.3 The standard penetration test. The driller raises the hammer and allows it to fall by alternately pulling and releasing the rope around the "cat's head."

regarding subsurface conditions is obtained in the field operation. Both thin-walled and split barrel samplers are shown in Figure 7.4.

7.1.3 Indirect Exploration Methods

Some subsurface explorations are made using geophysical techniques. These techniques involve determination of the properties of substrata in-directly by measuring either their electrical resistivity or the velocity at which a shock wave travels through them.

In the resistivity method (see Figure 7.5) two electrodes are inserted in the ground and connected to a current source. Intermediate electrodes are also inserted and connected by a potentiometer. When a reading is to be made, the potentiometer indicates the electrical resistivity of the ground within a depth equal to the electrode spacing a. The electrode spacing is then varied and more measurements are made. The resistivities so obtained are plotted versus depth or electrode spacing. Changes in slope of the resistivity versus depth plot indicate changes in subsurface materials. The resistivity measured is, of course, dependent on the amount of ionized salts and water present in the ground. An indication of the magnitudes of differences to be expected is given by Table 7.3.

Perhaps the most-used geophysical exploration technique with construction applications is the seismic refraction method. The essential features of

Figure 7.4 The Shelby tube thin-walled sampler and standard penetration or "split spoon" sampler (open) The heads of both samplers are provided with a ball check valve to allow water above the sample to escape, as the tube is filled, and to seal the tube on removal of the sampler from the test boring. The latter action results in better sample retention because of the partial vacuum developed.

the method are shown in Figure 7.6. A disturbance at the ground surface is produced by striking the ground with a sharp blow using a hammer or setting off an explosive charge. Geophones are used to determine the time taken for the disturbance to propagate several known distances from its source. These times are plotted versus distance, and changes in material are indicated by changes in the slope of this plot. The slope of the time-distance graph is the velocity of propagation of the disturbance. The velocity

TABLE 7.3   ELECTRICAL RESISTIVITIES OF SOILS AND ROCK

Material	Resistivity (ohm-centimeters)
Saturated organic clay or silt	500–2,000
Saturated inorganic clay or silt	1,000–5,000
Hard partially saturated clays and silts; saturated sands and gravels	5,000–15,000
Shales, dry clays, and silts	10,000–50,000
Sandstones, dry sands, and gravels	20,000–100,000
Sound crystalline rocks	100,000–1,000,000

Source: G. B. Sowers and G. F. Sowers, *Introductory Soil Mechanics and Foundations*, Macmillan, New York, 1970.

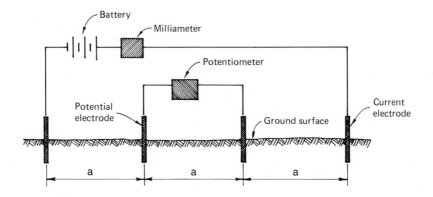

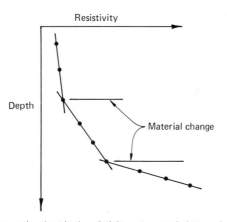

Figure 7.5 Schematic electrical resistivity setup and data reduction. (After Sowers and Sowers, 1970.)

and thickness of each stratum encountered can be computed. For example, the thickness of the upper stratum is given by Equation 7.1:

$$H_1 = \frac{A}{2} \sqrt{\frac{V_2 - V_1}{V_2 + V_1}} \tag{7.1}$$

In the case of seismic exploration, information of considerable utility is obtained, other than stratum boundaries. The wave propagation velocity determined for each stratum may be useful as an index of the resistance to excavation expected. Typical values of propagation velocities are given in Table 7.4. Various equipment companies have correlated seismic velocity with performance of tractor-ripper combinations. One such rippability chart is given in Figure 7.7. By knowing seismic propagation velocities and the rippability relationship for this particular equipment, a contractor may esti-

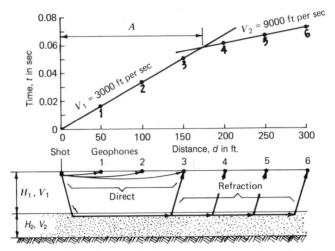

Figure 7.6 Subsurface exploration by seismic refraction. (After Sowers and Sowers, 1970.)

mate whether rock on a specific project may be ripped or would require shooting.

Indirect exploration methods are not widely used. In some cases, however, the results of these methods have been correlated directly with performance and utility of construction equipment and they may be extremely useful. But, it should be remembered that these methods rely on indirect measurements

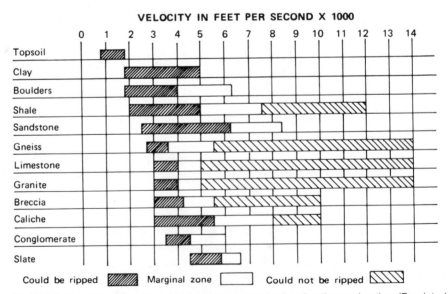

Figure 7.7 Rippability chart for excavation based on seismic refraction exploration. (Reprinted from *Engineering News-Record*, December 10, 1959, copyright © McGraw-Hill, Inc. All rights reserved.)

TABLE 7.4   WAVE VELOCITIES IN EARTH MATERIALS

Material	Velocity (ft/sec)
Loose dry sand	500–1500
Hard clay, partially saturated	2000–4000
Water, loose saturated soil	5200
Saturated soil, weathered rock	4000–10,000
Sound rock	7000–20,000

Source: G. B. Sowers and G. F. Sowers, *Introductory Soil Mechanics and Foundations*, Macmillan, New York, 1970.

of properties of subsurface materials and that these indirect measurements are not always reliable. Interpretation of geophysical exploration results is always best done by persons with a thorough background of experience with the method and equipment used. One may confidently use the results of geophysical exploration methods for engineering purposes if these tests are supplemented by tests using direct exploration or semidirect exploration techniques.

7.2 GROUNDWATER CONDITIONS

Explorations made using any of the techniques previously discussed can result in a reliable location of the groundwater table at the time of exploration. In many locations the position of the groundwater table is subject to variation throughout the year. These variations arise from many causes, among which are tides, pumping of wells, seasonal runoff variations, and changes in rainfall patterns. In some instances piezometers are installed in exploratory borings, and the fluctuations in the groundwater table are observed over a period of time. This information may not always be in the soils engineer's report. In many cases seasonal fluctuations are of sufficient magnitude to permit scheduling of construction to avoid the need for a dewatering system in excavations. When the matter of a few feet difference in groundwater elevation is important in planning construction operations, these seasonal variations and variations arising from other causes should be carefully considered.

In planning subsurface work it is important to consider not only the elevation of the groundwater surface but also the permeability of the soil. The design of a satisfactory dewatering system requires both, in addition to knowledge of location of the various strata. The determination of soil permeabilities may be accomplished in the laboratory or in the field as discussed in Chapter 5, according to the nature and geologic origin of the soils involved. Laboratory determinations are sufficient if soils are very uniform and good quality samples are obtained. Field tests that are properly planned and conducted are more universally applicable and reliable. These tests are done, as previously shown, by pumping from a test well and observing groundwater level changes at other locations, or by borehole permeability tests. Test conditions and observations and appropriate seepage analyses are combined to produce the desired parameters to quantify permeabilities.

Laboratory permeability tests on nonrepresentative samples can produce results that may be completely misleading. Ill-defined or poorly conducted pumping tests are equally as bad. In planning major dewatering operations, careful consideration of the completeness and validity of subsurface information is imperative.

7.3 SOILS ENGINEERING REPORTS

In preliminary planning for any major works the first step taken is often the site subsurface exploration. Following this exploration alternative designs for the project may be prepared. Results of these explorations usually form the sole basis for the contractor's interpretation of subsurface conditions to be encountered in construction. Soils reports are organized in a format similar to that shown below.

Scope and purpose
Introduction
Geologic setting
Field studies
Laboratory tests
Analysis
Conclusions and recommendations
Appendix

The subject headings are self-explanatory. Of these, the contractor will likely find that those sections related to site geology, field studies, and the appendix of information are the most useful. The geologic setting is of most value to the contractor who is familiar with conditions in the general area. Having worked in a similar geologic setting, he may fully appreciate the conditions that are to be encountered. The field study section will describe the exploration technique used and important observations made during the actual field work. The most definitive information is usually contained in the appendix, which will include a section showing drilling logs or soil profiles and laboratory test results. These may be separated or they may be included on single summary sheets, one for each boring. The drilling log and corresponding soil profile taken from a typical report are shown in Figures 7.8 and 7.9. The drilling logs will show the position of the groundwater table, standard penetration resistance test results, boring locations and elevations, and laboratory test results. In practically all cases it will be pointed out in the text of the report that the information shown on the boring logs is rep-

SOIL BORING RECORD

EXPLORATION FOR *Fernhill Center*					LOCATION *N.W. Building Corner see plot plan Elevation - 206*
DATE *28 March 1971*					
BORING NO. *B-3*					
RECORD BY *RES*					WATER TABLE *Measured at 15' depth 29 March 1973*
DRILL TYPE *4" Hollow stem*					
WEATHER *warm and sunny*					

SAMPLE NO.	TYPE TEST	DEPTH FROM	DEPTH TO	NO. OF BLOWS	SOIL DESCRIPTION AND BORING LOG
		0	4"		*asphaltic concrete*
		4"	12"		*gravel basecourse*
		1'			*light brown silty clay and clayey silt*
1-1	SPT*	5'	6.5'	3/3/5	*stiff light brown clayey silt with some mottling*
1-2	SPT	7.5'	9'	2/2/3	*same*
		10'			*driller notes softer drilling*
1-3	SPT	12.5'	14'	2/3/4	*same soil as above. top of sample very soft.*
1-4	SPT	17.5'	19'	3/10/12	*sample top same soil as above. bottom light gray clayey silt*
1-5	SPT	22.5'	24'	3/4/6	*light gray clayey silt*
1-6	SPT	27.5'	29'	4/7/7	*same*
		32'			*driller notes harder drilling*
1-7	SPT	32.5'	34'	4/11/13	*same soil but drier*
1-8	SPT	42.5'	44'	5/7/8	*same soil*
1-9	SPT	50'	50.6'	75/25-2"	*gravel and coarse sand. driller noted change at 46'*
REMARKS					*End boring at 50'* ** standard penetration test*

Figure 7.8 Example of drilling log or boring inspector's report.

resentative of conditions at the boring locations only. The assumption that conditions between borings are the same as conditions represented by the borings is one that may not always be warranted. By observing the relative uniformity of conditions at all borings and by knowing something of the geologic history of the particular site, one may better judge the reliability

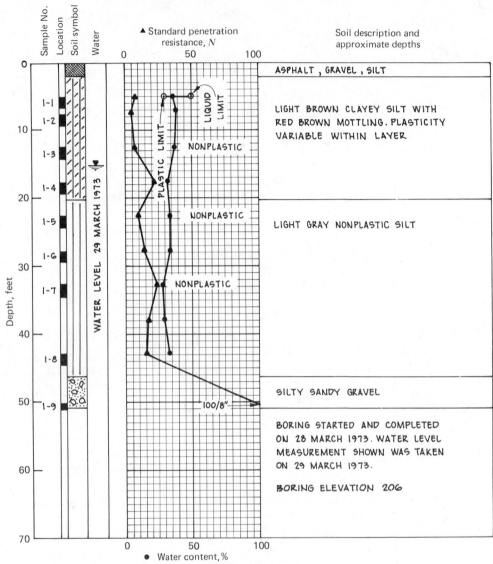

Figure 7.9 Subsurface profile constructed from drilling log and laboratory test results.

with which one may make such interpolations. It is prudent for the contractor to remember that, when he submits his bid, he usually must sign a statement to the effect that he fully understands subsurface conditions and their effects on the progress of his work (see Chapter 6). If conditions are not clear, he is well advised to assume that they are the worst that could be reasonably anticipated and to adjust his bid prices accordingly.

7.4   OTHER INFORMATION SOURCES

When for some reason available subsurface information is inadequate, the contractor may find it necessary to obtain his own. His choices range from seeking a full range of consulting services such as those described in earlier paragraphs, to seeking available information from other sources. This section discusses other possible sources.

Information on subsurface conditions from private sources is generally of limited availability. Private studies are the property of those who pay for them; hence, private consultants are obliged to release them only with the owner's permission. There is a wealth of information available from various public agencies, however. This information may be either general or specific in nature.

Certain general subsurface information is generated by federal and state agencies in the course of their normal duties. Such information may usually be obtained at a good library or from the agencies themselves. The U.S. Geological Survey (U.S.G.S.) publishes various maps, bulletins, and reports defining topography, geologic formations, and groundwater conditions of most sections of the country. This information is often of such large scale and low definition that it is of limited application for planning at a specific site. It is readily available, however, and should not be overlooked as a starting point for obtaining further information from other sources. Many state geologic agencies produce information similar to that provided by the U.S.G.S. In some cases states have produced reports and maps that define in great detail subsurface conditions on a small enough scale to provide reliable information at specific locations. This is particularly true in highly developed or rapidly growing areas. Groundwater records and well logs are always available from some governmental unit. Typically, a state engineer's or geologist's office maintains records on wells drilled in that state. These are usually filed by section, township, and range, or according to the owner's name at the time the well was drilled. Descriptions of soils, water levels, and bedrock locations contained therein are often useful. One should always be cautious in relying totally on such information, however, and consider its reliability in the light of how, by whom, and for what purpose it was obtained.

For specific site information, one may often obtain very detailed and valuable data from those governmental agencies having engineering or public works sections that have done work in a nearby area. The files of these agencies are public information, and those persons involved are generally cooperative and helpful in providing access. Practically no modern major civil works are executed without first obtaining some subsurface information. If these works are nearby or closely related to proposed work for which there is inadequate subsurface information, the contractor should seek out and make use of that which may be readily available as a result of earlier public work.

7.5 SUMMARY

Field explorations are usually made for design purposes. In only rare circumstances are they made at the direction of a contractor, who is reliant on the work of others and is usually obliged to bid only with the knowledge they have communicated. He is, therefore, obligated to acquire and maintain a reasonable degree of proficiency in interpreting soils engineering reports and to observe the effects of various subsurface conditions on his operations. With such a background he will be able to recognize adequate and inadequate information and thereby, one hopes, avoid the headaches arising from the need to argue for a claim of extra costs that result from changes in anticipated subsurface conditions.

After studying this chapter the reader should solve the accompanying problems. Having completed them, the reader should

1   Know what information to look for in a soils engineering report, and understand what it means.

2   Be able to locate existing subsurface information for a specific site when none has been provided.

REFERENCES

American Society for Testing and Materials, *1974 Book of ASTM Standards,* Philadelphia.

"How To Determine Rippability," *Engineering News Record,* December 10, 1959.

Sowers, G. B., and Sowers, G. F., *Introductory Soil Mechanics and Foundations,* Macmillan, New York, 1970.

Terzaghi, K., and Peck, R. B., *Soil Mechanics in Engineering Practice,* Wiley, New York, 1948.

PROBLEMS

1   Select a site in your locale, assuming that it is to be used for a borrow source. Further assume that the borrow site must produce 30,000 yd^3 of compacted fill. Prepare a map showing the site location and a plan view of the borrow area. From such public records as may be available and from your own observations and analysis, describe subsurface conditions at the site and any limitations that may be imposed on the borrow

operations by those conditions. Summarize your findings in a brief letter report or record memorandum.

2 For a local site of your own choosing, prepare a profile sheet indicating the sequence of subsurface materials you would expect to find to a depth of at least 50 ft. Describe the materials as completely as possible. Document the information on which your findings are based. Supplement your findings with personal observations during a site visit. A letter or memorandum should be prepared to summarize and transmit your findings.

CHAPTER 8

EMBANKMENT CONSTRUCTION AND CONTROL

Embankments are nearly always compacted when incorporated in engineered works. Some deep embankments are placed to full depth and compacted later. Compacting soil is a simple process. Energy in some form is applied to bring about densification. The process of densification results in the expulsion of air from the soil-water-air system. In the special case of saturated sand, only a two-phase system of high permeability is involved, and water is expelled from the soil voids. While the obvious effect of soil compaction is densification or an increase in unit weight, the purpose of compaction is to produce a soil mass with controlled engineering properties. Certain engineering properties of compacted soils are determined not only by density but also by the means and conditions under which the density was obtained.

Soil improvement through the application of additives during placement is usually intended to enhance engineering properties, but also may result in a change in the workability of the soil that will ease handling during construction.

Knowledge of the fundamentals of soil compaction and stabilization is essential to the understanding of specifications for these items of work in any contract, and particularly on large embankment projects like the dam shown in Figure 8.1. It will also greatly aid in the selection of equipment for actual construction.

This chapter deals with laboratory and field compaction techniques, methods of compaction control, and chemical stabilization of soils. The objectives of the presentation are to

1   Illustrate typical moisture-density relationships for coarse and fine soils, considering different levels of compactive effort.

2   Show the effects of various compaction conditions on engineering properties of fine and coarse soils.

3   Present standard laboratory compaction procedures.

4   Familiarize the reader with the intent, form, and content of the usual compaction specification.

5   Show how shrink and swell calculations should be made.

6   Indicate the appropriate use of specific types of compaction equipment and methods applicable to given soil conditions.

7   Explain the details of various methods for making and interpreting compaction control tests.

8   Introduce the reader to the fundamentals of chemical soil stabilization.

9   Discuss deep subsoil improvement by compaction and other means.

8.1 MOISTURE-DENSITY RELATIONSHIPS

The amount of densification obtained by any compaction process is dependent on the amount of energy used, the manner in which it is applied,

Figure 8.1 Construction of a major zoned embankment.

the type of soil involved, and the soil water content. This section discusses the relationships among these various factors.

The single-grained nature of the structure of cohesionless soils depends on gravity and the resulting grain-to-grain contact forces for its stability. If these contacts are disturbed, particle weight causes the structure to assume a new, more stable, and more dense configuration. It is important to note that the initial destruction of grain-to-grain contact will result in densification. On the other hand, application of a steady force to such a structure results primarily in an increase in the grain-to-grain forces with little permanent densification. For these reasons, it has been found that cohesionless soils are most efficiently compacted by vibrations, which disturb grain-to-grain contacts in the loose state.

Idealized moisture-density curves for a clean sand are shown in Figure 8.2. The curves are representative of the results of compacting the same soil using two different energy levels, E_1 and E_2. Several points on the figure are worthy of special note. Most obviously it is evident that greater energy input results in greater densification. But, logically, as higher energy levels are used, greater compaction results and further densification becomes increasingly difficult. Densification effects are, therefore, by no means proportional to the level of energy input. Also, the curves are S-shaped. That is, highest densities are obtained for a given compactive effort or energy level when the water content is either very low (approaching zero) or very high (approaching saturation). Thus, if a given density is specified, say that represented by the horizontal dashed line, it may be obtained in one of several ways. The possibilities are indicated by points a, b, and c. For instance, compactive effort will be minimized if moisture conditions are adjusted to be either very dry or very wet (a or b) More compactive energy

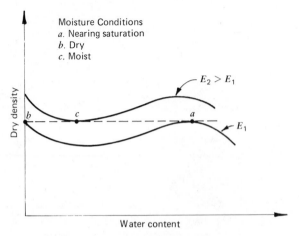

Figure 8.2 Idealized moisture-density relationships for clean sand.

will be required for moist conditions (*c*). The choice in practice is dictated by the economics of providing moisture control or compaction energy. The characteristic *S*-shape of the curves is the result of surface tension in the soil moisture that results in apparent cohesive strength, as explained in Chapter 5 and, therefore, resistance to compaction. When the soil is dry or saturated, these effects do not exist and greatest density for a given energy level is achieved. When the soil is moist, capillary forces impart strength (apparent cohesion) to the soil, which results in increased interparticle forces and reduced densification.

Moisture-density curves characteristic of a cohesive soil are shown in Figure 8.3. As in the case of cohesionless soils, increased compactive energy is seen to result in increased densification. Curves representing the different compactive efforts are of similar shape. Maximum densification for a given effort is obtained at some intermediate moisture content. The points *a* and *b* are referred to as having the coordinates *maximum dry density* and *optimum water content* for a given compactive effort. It is believed that this optimum condition results from resistance to densification, on the dry side by forces in tightly held soil water, and on the wet side by the high degree of saturation and low permeability of the mass. It is apparent that a specified density such as that at *c* may be obtained with widely different compactive efforts and under widely different moisture conditions. The possibilities in this case are indicated by points *a, d,* and *f* for the two compactive efforts shown. The proper choice of moisture content is dictated by the economics of moisture control and available compaction energy. If, as is often the case, a range of moisture contents is specified along with minimum density, the range of available choices of compactive efforts is narrowed.

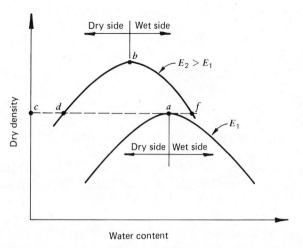

Figure 8.3 Idealized moisture-density relationships for silty clay.

8.2 PROPERTIES OF COMPACTED SOILS

Engineering properties of cohesionless soils are primarily a function of relative density. Relative density D_r may be defined in terms of void ratio:

$$D_r = \frac{e_{(max)} - e}{e_{(max)} - e_{(min)}} \times 100 \qquad (8.1)$$

or in terms of unit weights:

$$D_r = \frac{\gamma_{d(max)}}{\gamma_d} \frac{\gamma_d - \gamma_{d(min)}}{\gamma_{d(max)} - \gamma_{d(min)}} \times 100 \qquad (8.2)$$

The following definitions apply:

e	= void ratio measured
$e_{(max)}$	= maximum void ratio determined by standard test method
$e_{(min)}$	= minimum void ratio determined by standard test method
γ_d	= dry density measured
$\gamma_{d(max)}$	= maximum dry density determined by standard test method
$\gamma_{d(min)}$	= minimum dry density determined by standard test method

As relative density increases, soil strength increases and compressibility decreases. Permeability is reduced. It will be recalled that cohesionless soil structure is single grained. With this type of structure there is usually no important difference between a natural deposit or a fill, provided both are at the same relative density. Since the objective of compaction is to produce a fill with controlled properties, usually only density is specified for these types of soils. Seldom in compaction specifications will control of water content be specified. In those cases where water control is specified, it is usually done to ensure maximum compaction for the energy available.

Engineering properties of plastic soils are very much affected by compaction water content. Water content during placement influences the structure of clay soils sufficiently to have a significant effect on strength, permeability, and compressibility. Studies of compacted soils at the microlevel have shown that soils compacted on the dry side of optimum moisture content have a flocculent structural arrangement of particles. On the wet side of optimum moisture content, the structure is dispersed. A schematic representation of these findings is shown in Figure 8.4. Points a and b are at the same density. Soil structure influences engineering properties. Compaction water content must, therefore, be controlled along with density if the influence of structure is of importance in the design of an embankment.

Wet side as opposed to dry side compaction, for instance, may reduce the coefficient of permeability of a soil dramatically, as shown in Figure 8.5, even though density is constant. The larger pore sizes in the flocculent

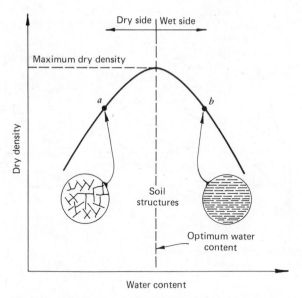

Figure 8.4 Effects of water content during compaction on soil structure. (After Lambe, 1958.)

structure, even though few in number, permit more flow than a larger number of smaller openings in the dispersed structure. Where permeability control, notably in the core section of earth dams, is important, a specification for wet side compaction may be made.

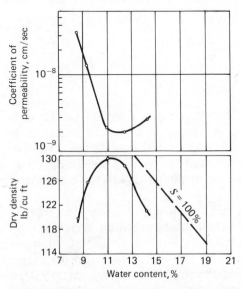

Figure 8.5 Effects of water content during compaction on soil permeability. (After Lambe, 1958.)

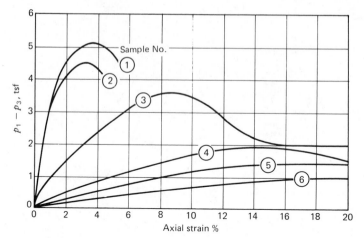

Figure 8.6 Stress-strain relationships for compacted clay soil. (After Seed and Chan, 1959.)

Stress-strain curves from triaxial shear tests on soils compacted at various water contents are shown in Figure 8.6. Wet side compaction (Sample Nos. 4, 5, 6) yields a flexible material that has low strength. Wet side compaction of an earth dam core produces an impervious barrier to water flow, which can yield without cracking as the dam deforms. Its low strength must be compensated for by stronger construction in the outer zones or shells of the structure. Dry side compaction (Sample Nos. 1 and 2) yields a structure that is brittle but of high strength.

8.3 LABORATORY COMPACTION PROCEDURES

Standardized laboratory procedures are needed to clearly specify construction requirements. They also are useful in studies related to the effects of compaction techniques on soil properties. Different agencies use slightly different methods, but all methods are similar. They specify the soil to be used and compactive effort to be exerted in developing the moisture-density relationship.

For plastic soils the most widely used test method is the standard compaction (ASTM D698) test procedure, shown in Figure 8.7. It is sometimes referred to as the Standard Proctor test, after the pioneer of these methods of control. The Standard Proctor test, however, is slightly different from ASTM D698. The modified compaction test (ASTM D1557) was developed to permit compaction control where very high densities are to be specified. The standard and modified test methods are summarized in Table 8.1.

Figure 8.7 The standard laboratory compaction test.

Note therein that the modified procedure incorporates a significantly higher energy level than the standard test. Its use will result in significantly higher density than the standard test at the same water content. For most soils the maximum dry density from the standard test is 90 to 95 percent of the

TABLE 8.1   COMPARISON OF COMPACTION TEST PROCEDURES

Designation	Standard ASTM D698	Modified ASTM D1557
Mold		
Diameter (in.)	4	4
Height (in.)	4 5/8	4 5/8
Volume (ft^3)	1/30	1/30
Tamper		
Weight (lb)	5.5	10.0
Free drop (in.)	12	18
Face diameter (in.)	2	2
Face area (in.^{2})	3.1	3.1
Layers		
Number, total	3	5
Surface area, each (in.^{2})	12.6	12.6
Compacted thickness, each (in.)	1 5/8	1
Effort		
Tamper blows per layer	25	25
(ft-lb/ft^3)	12,375	56,250

corresponding value from the modified test. The results of a standard compaction test are presented in Figure 8.8. The zero air voids line (coordinates of water content and dry density for the saturated condition, Equation 8.3) is shown for reference.

$$\gamma_{d(zav)} = \frac{G_s \gamma_w}{1 + wG_s} \qquad (8.3)$$

For cohesionless soils, special laboratory compaction techniques are necessary. Tests using the basic procedures outlined in Table 8.1 are sometimes specified, although often inappropriately. It has become more common in recent years to use relative density control for these materials. Tests for maximum and minimum density or void ratio are, therefore, needed. These tests (ASTM D2049) employ loose pouring into an open container to determine minimum density, and vibration of the container and contents to determine maximum density. It should be noted that relative density values are very sensitive to slight changes in the laboratory values of maximum and minimum density (see Equation 8.2). These values must, therefore, always be determined by standard methods if relative density is to be used as a construction control index.

8.4 COMPACTION SPECIFICATIONS

The special specifications for embankments in the construction contract may be one of two general types. A work-type specification tells the contractor what to do and how to do it. If he complies, the owner is obligated to accept the result. This type of specification should produce the lowest of bids, for the contractor knows exactly what must be done to accomplish the work. A performance specification, on the other hand, requires that the contractor achieve a specific end result. Thus, the contractor is told what he must accomplish. His method of doing it may require some experimentation on his part and his bid might be correspondingly higher. For either type of specification, compaction control is ultimately based on a comparison of field results with laboratory tests. For cohesionless soils, relative density (Equation 8.1 or 8.2) is usually specified. For cohesive soils relative compaction

$$RC = \frac{\gamma_d}{\gamma_{d(max)}} \times 100 \qquad (8.4)$$

and water content are specified. Relative compaction specifications are sometimes used for cohesionless soils as well.

To develop a work specification the contractor builds a test fill and the adequacy of his compaction method is established. Alternatively, through

MOISTURE – DENSITY RELATIONSHIP

TEST FOR _Lone Oak Road Embankment_

TEST BY _FJB_   DATE _June 26, 1973_

SAMPLE IDENTIFICATION _Red-brown Silty Clay_

COMPACTIVE EFFORT _ASTM D698_

TRIAL NUMBER	1	2	3	4	5	6
Wt. of Wet Soil & Cylinder	12.92	13.12	13.32	13.35	13.28	
Wt. of Cylinder	9.66	9.66	9.66	9.66	9.66	
Wt. of Wet Soil	3.26	3.46	3.66	3.69	3.62	
Vol. of Cylinder	1/30	1/30	1/30	1/30	1/30	
Unit Wet Weight	97.9	103.9	109.9	110.8	108.7	
Unit Dry Weight	81.5	84.1	86.1	83.4	79.4	
Pan Number	161	181	189	185	170	
Wt. of Wet Soil & Pan	80.58	92.92	88.92	101.65	93.45	
Wt. of Dry Soil & Pan	72.57	81.43	76.68	84.49	76.99	
Wt. of Water	8.01	11.49	12.24	17.16	16.46	
Wt. of Dish	32.73	32.46	32.44	32.37	32.40	
Wt. of Dry Soil	39.84	48.97	44.24	52.12	44.59	
Water Content, %	20.1	23.5	27.7	32.9	36.9	

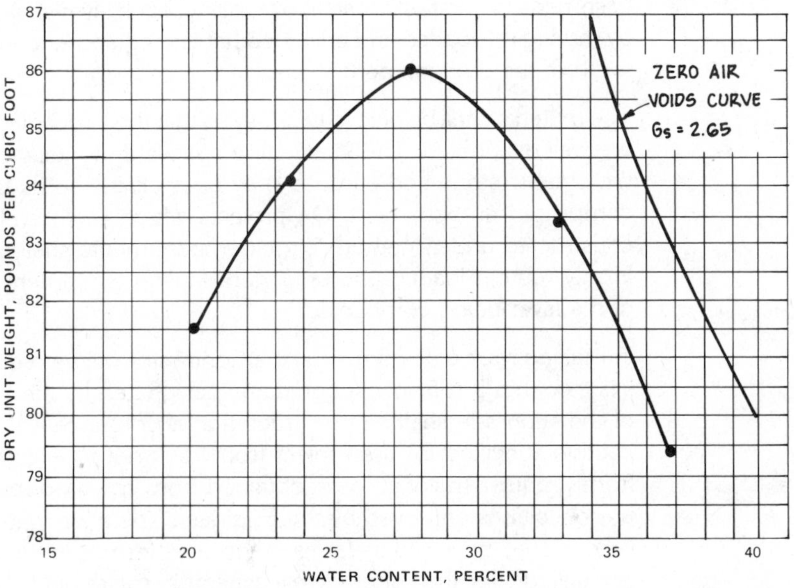

Figure 8.8 Standard laboratory compaction test results.

his experience with a particular soil, the engineer may know what is required to achieve adequate compaction. The specification in either case is written after direct or previous field experience has demonstrated the sufficiency of the compaction method to be specified. An example of a work-type specification is given below.

2-03.3(14)A Rock Embankment Construction.

Rock embankments shall be constructed in horizontal layers not exceeding eighteen (18) in. in depth, except that when the average size of the rocks exceeds eighteen (18) inches, the layers may be as deep as required to allow their placement. Occasional rocks exceeding the average size may be disposed of as approved by the Engineer instead of being incorporated in the embankment.

Each layer shall be compacted with at least one full coverage with a 50-ton compression type roller or four full coverages with a 10-ton compression type roller for each 6-in. depth of layer or fraction thereof. The number of coverages for compression type rollers, including grid rollers, weighing more than 10 tons and less than 50 tons shall be as directed by the Engineer. Rollers shall be so constructed that they will exert a reasonably uniform pressure over the area covered. In lieu of the foregoing, each layer shall be compacted with four full coverages with an approved vibratory roller for each 6-in. depth of layer or fraction thereof. Rolling may be omitted on any layer or portion thereof only when in the judgment of the Engineer it is physically impractical of accomplishment. In addition to the above rolling, each layer shall be further compacted by routing the loaded and unloaded hauling equipment uniformly over the entire width of the embankment.

The material shall be placed carefully so that the larger pieces of rock or boulders are well distributed. The intervening spaces and interstices shall be filled with the smaller stone and earth as may be available so as to form a dense, well-compacted embankment. On projects where earth embankments are to be constructed under Method A, rock embankments shall be compacted by uniformly routing loaded and unloaded hauling equipment over the entire width of the layer being compacted.

In making rock embankments, the Contractor will be required to bring up such fills to within 2 ft below subgrade as designated by the Engineer. The top 2 ft of the subgrade shall be constructed in accordance with the method of compaction specified for the project from rock not to exceed 4 in. in size and/or from granular material to be obtained from the roadway excavations or from borrow pits as approved by the Engineer. The finer materials from rock excavations shall be saved as far as practicable for use in topping out rock fills and backfilling over the subgrade excavation in rock cuts.

When it is specified that the subgrade shall be trimmed with an automatically controlled machine, no rocks or stones larger than 2 in. shall be left within 4 in. of the subgrade. (From *Standard Specifications for Road and Bridge Construction,* State of Washington Department of Highways, 1963.)

In a performance specification, the contractor is told to achieve a specified relative density or relative compaction and that tests will be conducted during construction to ensure compliance. The performance specification is the more common of the two types. It is illustrated below, under **Methods B and C**.

2-03.3(14)B   Earth Embankment Construction.   Earth embankments shall be constructed in compacted layers of uniform thickness by one of the three methods, A, B, and C, described in subsequent sections. Under all methods the layers shall be carried up full width from the bottom of the embankment. The slopes of all embankments shall be compacted to required density as part of the embankment compaction work.

Unless otherwise stated in the Special Provisions, earth embankments shall be constructed by Method B.

During the grading operations, the surface of embankments and excavations shall be shaped to a uniform cross section. All ruts and depressions capable of ponding water shall be eliminated.

On tangents, the center of embankment layers shall be constructed higher than the sides. Side hill embankments shall be constructed with the intersection with the original ground as the high point of the layer and shall uniformly slope to the outer side with a slope not to exceed 1 ft in 20 ft.

2-03.3(14)C   Compacting Earth Embankments.   Except when Method A is specified, earth embankments shall be compacted with modern, efficient, compacting units satisfactory to the Engineer. The compacting units may be of any type provided they are capable of compacting each lift of the material to the specified density. The right is reserved for the Engineer to order the use of any particular compacting unit discontinued if it is not capable of compacting the material to the required density in a reasonable time.

The determination of field in-place density shall be made in accordance with methods and procedures approved by the Engineer.

Method A.   Under Method A, earth embankments shall be compacted in successive horizontal layers not exceeding two (2) feet in thickness, and each layer shall be compacted by routing the loaded haul equipment over the entire width of the layer. When permitted by the Engineer, side hill fills too narrow to accommodate the hauling equipment may be placed by end dumping until the embankment material can be spread to sufficient width to permit the use of hauling equipment upon it. Thereafter the remainder of the embankment shall be placed in layers and compacted as above specified. Suitable small mechanical or vibratory compactor units shall be used to compact the layers adjacent to structures that are inaccessible to the loaded haul equipment.

Method B.   Under this method each layer of the top 2 ft of embankments shall be compacted to 95 percent and each layer of embankments below the top 2 ft shall be compacted to 90 percent of the maximum density as determined by compaction control tests specified in Section 2-03.3(14)D. Horizontal layers shall not exceed 8 in. in loose thickness except that the layers of the top 2 ft shall not exceed 4 in. in loose thickness. Moisture content of the embankment material at the time of compaction shall be as specified in Section 2-03.3(14).

At all locations that are inaccessible to compaction rollers the embankments shall be compacted in layers as required herein and shall be compacted to the required density by the use of small mechanical or vibratory compactor units.

Method C.   Under this method each layer of the entire embankment shall be compacted to 95 percent of the maximum density as determined by compaction control tests specified in Section 2-03.3(14)D. The horizontal layers shall not exceed 8 in. in loose thickness except that the layers of the top 2 ft shall not exceed 4 in. in loose thickness. The moisture content of the embankment material at the time of compaction shall be as specified in Section 2-03.3(14).

At all locations that are inaccessible to compaction rollers the embankment shall be constructed in layers as required herein and shall be compacted to the required density by use of small mechanical or vibratory compactor units. (From *Standard Specifications for Road and Bridge Construction,* State of Washington, Department of Highways, 1963.)

The distinction between relative density and relative compaction is not always clear. The two concepts are compared graphically in Figure 8.9. It should be noted that it is possible for field relative density to be either less than zero percent or greater than 100 percent, and that relative compaction

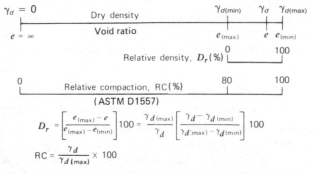

Figure 8.9 Relative compaction-relative density relationships. (After Lee and Singh, *9th Annual Engineering Geology and Soils Engineering Symposium*, Boise, Idaho, 1971.)

may exceed 100 percent. Tests have shown that, in a large number of cases involving sands and gravel, minimum density corresponds to about 80 percent relative compaction based on the ASTM D1557 method.

Having established water contents and densities corresponding to the engineering properties desired in the fill, the engineer prepares a compaction specification. The essence of the specification is that a certain percentage of relative density or relative compaction will be required for specific soils. Water content may also be specified if compaction water content influences the fill properties. For example, the U.S. Bureau of Reclamation requires compaction along the lines indicated in Table 8.2 for dams less than 50 ft high.

The specification in Table 8.2 for cohesive soils recognizes the need to control both density and water content. The specification for sands recognizes the importance of density and the need for nearly full saturation during construction to achieve it. Both specifications become more restrictive as dam height increases and the need for controlling engineering properties becomes more important.

8.5 SHRINK AND SWELL FROM BORROW TO FILL

Soil materials are usually placed in a compacted fill at greater density than that in their natural state in the borrow area. This is not always true, but it usually is. Intact rock usually is denser in its natural state than in a fill constructed with the rock after it has been ripped or shot. The amount of shrink or swell from borrow to fill is important, because payment for fill is usually based on the fill volume quantity, yet cost of doing the work is affected by the excavation and haul quantities, which are different. The quantities

TABLE 8.2   EXAMPLE OF EARTHWORK COMPACTION REQUIREMENTS FOR EARTH DAMS

Compaction Requirements	Water Control
Cohesive Soils	Optimum ± 2 percent
0–25 percent retained on No. 4; RC = 95 percent	
26–50 percent retained on No. 4; RC = 92.5 percent	
More than 50 percent retained on No. 4; RC = 90 percent	
Cohesionless Soils	Very wet
Fine sand	
0–25 percent retained on No. 4; $D_r = 75$	
Medium sand	
0–25 percent retained on No. 4; $D_r = 70$	
Coarse sand and gravel	
0–100 percent retained on No. 4; $D_r = 65$	

during these operations can be computed rationally, using information from subsurface exploration reports.

Consider a case where the subsurface exploration has shown that the soil dry density and water content in a borrow area for a job are 83 pcf and 20.2 percent, respectively, and that a 100,000 yd³ fill must be built under a specification that calls for minimum relative compaction of 95 percent. Test results in the soils engineer's report show that the maximum dry density of the soil is 112 pcf. The specifications therefore require that the minimum density of the fill be 106.4 pcf. In order to produce 1 ft³ of fill, at least 106.4/ 83 or 1.28 ft³ of borrow will be required. In going from borrow to fill, therefore, the shrinkage is

$$\frac{1.28 - 1}{1.28} \times 100 = 22 \text{ percent}$$

of the borrow volume, or

$$\frac{1.28 - 1}{1} \times 100 = 28 \text{ percent}$$

of the fill volume. Since the estimate is usually based on fill volume for a fill project, the appropriate shrinkage factor is the latter of these two numbers. In the case of our assumed project the excavation quantity would be 28 percent greater than the fill volume, or 128,000 yd³. If the project had been a 100,000 yd³ excavation project and the problem was to find a suitable disposal area for placement and compaction of the excavation spoil, the appropriate shrinkage would have been 22 percent. That is, 100,000 yd³ of excavation would produce 78,000 yd³ of fill.

This example illustrates how shrinkage calculations are appropriately made. It also demonstrates how tabulated shrinkage values taken from the literature may be seriously misused if they are not clearly defined. In the example, depending on the definition of shrinkage, an error of about 6 percent could be made in estimating a job if the wrong factor were chosen. This seemingly small percentage could be very important in terms of dollar cost on a large job. Of course, shrinkage estimates should never be based on unsubstantiated numbers from the literature if site specific information is available.

In the foregoing example, it was shown that 128,000 yd³ of excavation would be required to produce 100,000 yd³ of fill. The haul estimate might be based on yardage or tonnage. If it is to be based on yardage, then a swell factor from borrow to the haul unit will have to be derived. There is no rational way of doing this, except by quantifying one's experience. The same problem, however, may be solved rationally if the tonnage rating of the haul units is known. For purposes of this illustration, if it is assumed that the haul unit capacity is 50 tons, the number of trips required for the haul is about

$$\frac{128,000 \times 27 \times 83 \times 1.202}{2000 \times 50} = 3448 \text{ trips}$$

8.6 COMPACTION EQUIPMENT FOR SHALLOW LIFTS

To meet compaction specifications the contractor must select the proper equipment to do the job, and use it correctly. It is the engineer's responsibility to see that the specifications are met.

Selection of equipment is keyed to the nature of soil to be compacted, the degree of compaction required, and the space available in which to do the job. It is well known that sands are most efficiently compacted by vibration while cohesive soils are better compacted by pressure that is maintained for some time. This is logical, considering earlier discussions of soil structure and plasticity. Some indication of the acceptability of various large compaction equipment is given by Figure 8.10. Among the various types of compactors available, there are many variations of size and shape of the compacting element. Figures 8.11, 8.12, and 8.13 show several of the types of rollers indicated in Figure 8.10.

It would be desirable, in advance of construction, to know how the properties of the soil to be placed relate to the optimum conditions for compaction. This information provides a realistic basis for selecting equipment and scheduling to allow for additional watering or drying of embankment materials. In general, it is found that the shape and position of the moisture-

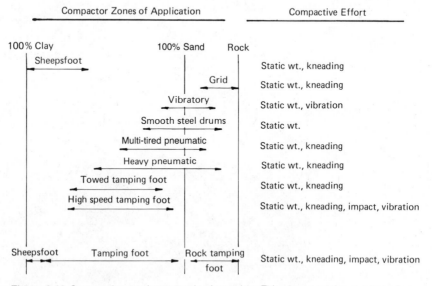

Figure 8.10 Compaction equipment selection guide. This chart contains a range of material mixtures from 100 percent clay to 100 percent sand, plus a rock zone. Each roller type has been positioned in what is considered to be its most effective and economical zone of application. However, it is not uncommon to find them working out of their zones. Exact positioning of the zones can vary with differing material conditions.

Figure 8.11 Compaction equipment for coarse materials. *(a)* Smooth drum vibratory roller for compacting granular materials. *(b)* A grid roller for compacting rock.

density curve for cohesive soils depend on soil plasticity characteristics. These effects are illustrated by Figure 8.14. The more plastic materials do not compact to high densities. The densities obtained for them are fairly insensitive to moisture content changes. The soils of lower plasticity must be compacted within fairly narrow ranges of water content to attain optimum conditions. Figure 8.15 relates optimum water content for a large number

Figure 8.12 A light pneumatic roller.

Figure 8.13 A self-propelled tamping or "sheepsfoot" roller.

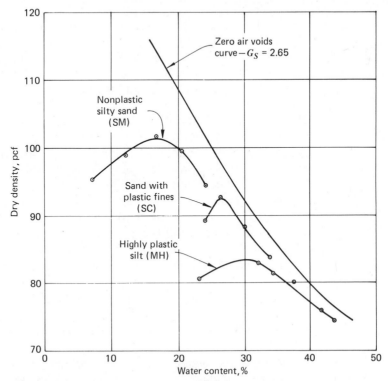

Figure 8.14 Moisture-density curves indicating the effects of soil plasticity differences.

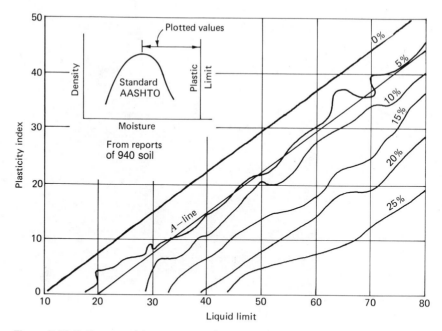

Figure 8.15 Optimum moisture content related to soil plasticity. (After McDonald, *10th Annual Engineering Geology and Soils Engineering Symposium*, Moscow, Idaho, 1972.)

of soils to their plasticity characteristics. Figure 8.16 empirically relates density at optimum water content to optimum water content. Using Figure 8.15, the soil's optimum water content may be estimated using classification indices. This may be compared with field water content in a borrow area to assess the workability of the material. Figures 8.15 and 8.16 may be used with field test data to make preliminary estimates of material quantities.

Suppose that laboratory tests on soil to be used in a fill show that its liquid limit and plastic limit are 46 and 30, respectively. Its natural water content is about 28 percent and its dry density, in the borrow area, is about 86 pcf. Figure 8.15 indicates that the optimum water content for this soil (an ML in the Unified Classification System) is about 10 percent below its plastic limit, or about 20 percent. We learn, in addition, from Figure 8.16 that the wet density of the soil at its optimum water content is about 122 pcf. Its dry density at optimum water content would thus be about (See Equation 3.4)

$$\frac{122}{1.20} = 101.7 \text{ pcf}$$

From the foregoing illustration we have learned that the soil will be very wet of optimum water content and will require drying for efficient placement. Furthermore, excavation will possibly be difficult. Assuming that the compaction specification calls for 95 percent of standard maximum dry density, or

$$0.95 \times 101.7 = 96.6 \text{ pcf}$$

shrinkage from borrow to fill should be about

$$\frac{96.6 - 86}{96.6} \times 100 = 11 \text{ percent}$$

of the fill volume.

8.7 COMPACTION CONTROL TESTING

The ultimate purpose of compaction control efforts is to ensure that the owner in the contractual relationship receives a fill that will serve his purpose. The engineer will sample the fill periodically to ensure compliance with the specifications. The fill may be checked after a certain number of cubic yards has been placed, after a new lift has been placed, or sometimes after each shift. On large jobs the contractor's operation may sometimes be made more efficient by careful interpretation of these test results.

Compaction control tests involve determination of the in-place density and water content of the fill. A number of methods are available. Field density is determined either directly or indirectly. Conventional tests such as the sand cone (ASTM D1556), Washington Densometer (ASTM D2167), or oil

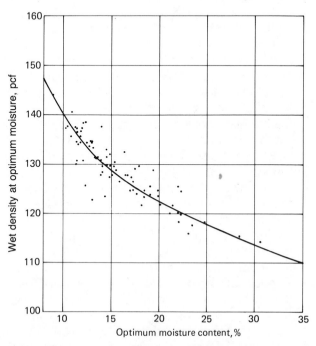

Figure 8.16 Empirical relationship of optimum moisture content and wet density at optimum moisture content. (After Hilf, American Society for Testing and Materials, STP No. 232, 1957.)

replacement method involve excavation of a small hole, the material from which is weighed. The volume of the hole is found by filling it with sand of a known density, a balloon full of water, or oil. The weight of soil to volume of soil ratio is the wet density. Water content of the soil removed is also determined and the dry density is calculated using Equation 3.4. The rapid determination of moisture content is essential so that the contractor will have his fill approved or disapproved before placing more material. Figure 8.17 illustrates the results for the sand cone test. A test setup is shown in Figure 8.18a. It should be noted that the usual procedure in the test involves compacting one cylinder of the soil removed to ensure that it fits the standard moisture-density curve being used as a reference for relative compaction calculations.

In the right kinds of soils, undisturbed samples of regular shape may be taken by sampling tubes or hand trimming. This permits a volume determination to be made without replacement by some other substance. It is only necessary to calculate volume directly. An indirect method that employs radioactive isotopes, however, is more convenient and rapid, especially on larger jobs, than this or the other methods discussed above.

Nuclear density gages (see Figure 8.18b) consist of two essential parts, a radiation source and a radiation counter. The source is placed either on the fill surface or at some depth. To make the density determination, a gamma radiation source is used. The emissions sensed by the counter are inversely proportional to the density of the material (soil) through which they pass. The count rate is, thus, calibrated to reflect this property. For a water content determination, a high-energy neutron source is used. The emissions are reflected and counted at a rate proportional to the soil water content. With proper calibration the method gives accurate results. Its principal benefits, however, are that the results are available immediately and that large numbers of tests may be conveniently made.

The standard compaction test (ASTM D698) is run on material passing either the No. 4 sieve or the 3/4-in. sieve. Any larger material is removed. Yet field density tests involve samples that contain this fraction, which is not included in establishing the laboratory standards for the project. What effect does this have on the compliance with specifications? The following example illustrates the point. The procedure is a method commonly used for correction for the oversize fraction.

Assume a field density test has been run on a soil with 11 percent of its weight retained on the No. 4 sieve. The test showed that the dry density was 124 pcf and that the water content was 12 percent for the total sample. The specifications, based on ASTM D698, require a minimum dry density of 122 pcf and a maximum water content of 12 percent. Were they satisfied? Figure 8.19 shows a breakdown of the constituents, using a specific gravity of 2.64 for the solids. The assumption must be made in correcting for oversize material that the water is associated with a certain size range of the solids in some regular fashion. Commonly, it is assumed that all of the moisture is

FIELD DENSITY — SAND CONE METHOD

TEST FOR _Site Fill - Test 3_

TEST BY _TKN_ DATE _6 JUNE 1973_

SOIL DESCRIPTION _gravelly sand_

TEST LOCATION _see sketch_

FIELD DENSITY		CYLINDER	
1. Wt. Sand + Jar	_11.90_	1. ASTM Designation	_D 1557_
2. Wt. Residue + Jar	_5.78_	2. Wt. Soil + Mold	_9.63_
3. Wt. Sand Used (1) - (2)	_6.12_	3. Wt. Mold	_9.36_
4. Wt. Sand in Cone & Plate	_3.41_	4. Wt. Soil	_4.74_
5. Wt. Sand in Hole (3) - (4)	_2.71_	5. Wet Density, pcf (4) ÷ Vol. Mold	_142.2_
6. Density of Sand	_89.5_		
7. Wt. Container + Soil	_4.57_	6. Moisture Content	_6.76_
8. Tare Wt. Container	_0.16_	7. Dry Density, pcf (5) ÷ [1 + (6)]	_133.2_
9. Wt. Soil (7) - (8)	_4.41_		

MOISTURE CONTENT — PAN NO. _110_

WW _127.25_   DW _121.30_

DW _121.30_   TW _33.25_

WATER _5.95_   SOIL _88.05_

PER CENT MOISTURE _6.76_

FIELD DENSITY	
10. Vol. of Hole (5) ÷ (6)	_0.0303_
11. Wet Density, pcf (9) ÷ (10)	_145.54_
12. Moisture Content, %	_6.56_
13. Dry Density, pcf (11) ÷ [1 + (12)]	_136.58_
Representative Curve No.	_2_
14. Optimum Moisture, %	_7.0_
15. Maximum Dry Density, pcf	_133.3_
16. Relative Compaction, % (13) ÷ (15)	_102.0_

FIELD MOISTURE DETERMINATION

Method _oven_

Percent Moisture _6.56_

NOTES:

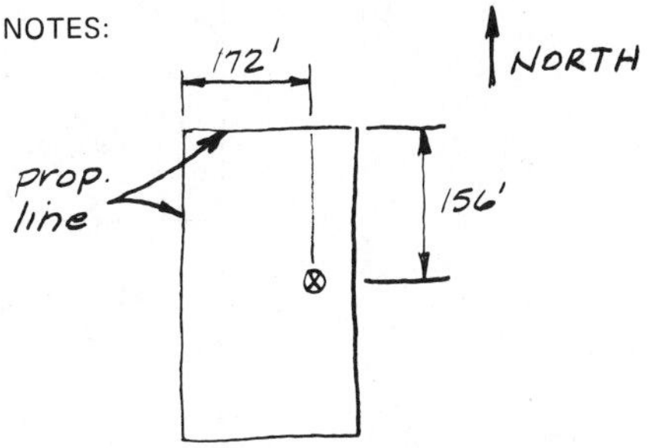

Test @ EL. 26.6
depth in fill = 18"

Figure 8.17 Field density test results using the sand cone method.

Figure 8.18 Common methods of field density testing. *(a)* The sand cone test. *(b)* Nuclear density testing.

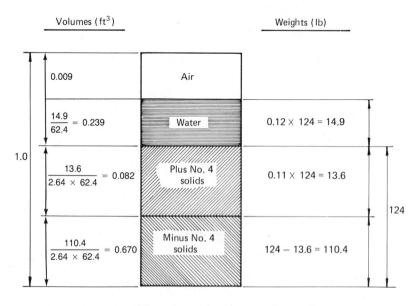

Figure 8.19 Phase diagram illustrating an example of the oversize correction.

held by the minus No. 4 material. Alternatively, the moisture content of the coarser fraction may be measured by laboratory tests. For purposes of this illustration, if the water content of the plus No. 4 soil is taken as zero, for the minus No. 4 material,

$$w = \frac{14.9}{110.4} \times 100 = 13.5 \text{ percent}$$

and the dry density is

$$\gamma_d = \frac{110.4}{0.918} = 120.3 \text{ pcf}$$

Therefore, the soil fraction for which the specification was developed is wetter than allowed and has not been sufficiently compacted. The specification is not met, yet, for the total sample, it appeared that it had been.

The following example illustrates how field and laboratory test results may be used to modify an unsatisfactory compaction operation. It should be remembered that to change compaction results we may change compaction energy or water content or both. Consider Figure 8.20. It has been experimentally demonstrated that a semilogarithmic plot of maximum dry density versus compactive effort, such as that shown in Figure 8.20b, is linear. Assume that the soil for which these results were obtained is to be compacted to 100 percent of maximum density according to ASTM D1557 and at optimum water content. Further assume that the project has begun and that a field density test shows that the fill has a dry density of 106 pcf and a water content of 14 percent. How should the compaction procedure be changed to produce the desired result? In Figure 8.20a the coordinates representing the fill conditions have been plotted, and the compaction curve corresponding to the compactive effort being used has been sketched. For this curve the maximum dry density for the field compactive effort is about 109 pcf. From Figure 8.20b the field compactive effort may be estimated by intersecting the laboratory curve at this density. Since the field effort is equivalent to about 20,000 ft-lb/ft^3 and the effort in the ASTM D1557 test on which the specification is based is about 56,000 ft-lb/ft^3, the field effort will have to be increased by about 2.5 times to achieve the desired results. Thus, if the roller has been making two passes over the fill, the procedure should be changed so that it makes at least five passes before compaction is checked again. Note that in this case water content should also be increased about 1 percent to comply with specifications. It is of course possible that the roller being used might not be capable of obtaining the required density with the lift thickness being used.

The foregoing paragraphs have considered compaction in shallow lifts on the order of 2 ft thick or less. Surface rolling is not usually effective below this depth. Figure 8.21a illustrates the depth-density relationship resulting from vibratory compaction of a cohesionless material. Maximum compaction usually occurs a foot or two below the ground surface. To achieve some

	LAYERS	BLOWS PER LAYER	WT. OF HAMMER	DROP IN INCHES	COMPACTIVE EFFORT FT-LB/CU FT
⊗	5	55	10 LB	18	55,900 ASTM D1557
○	5	26	10 LB	18	26,400
□	5	12	10 LB	18	12,200 ASTM D698

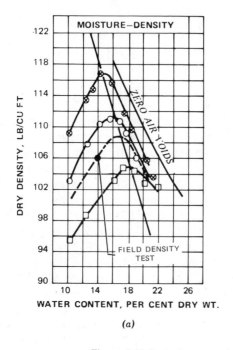

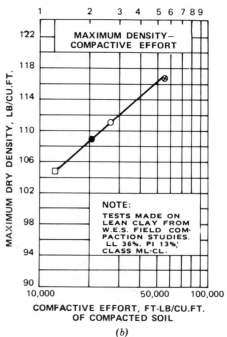

Figure 8.20 Illustrative examples of modification of compaction procedure based on laboratory tests and field density tests. (Adapted from *Soil Manual* (MS-10), published by the Asphalt Institute.)

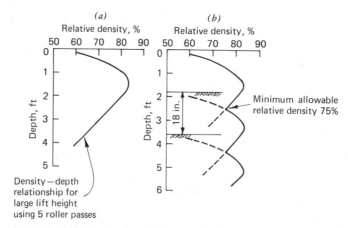

Figure 8.21 Depth-density relationships for compaction of a granular soil by vibratory roller. (After D'Appolonia, Whitman, and D'Appolonia, 1969.) *(a)* Single pass. *(b)* Multiple lifts.

specified density then, the lift thickness should not exceed the depth to the minimum acceptable density on the plot. Figure 8.21*b* shows the compaction achieved in a number of lifts, illustrating the need for lift thickness control. In a cohesive soil, compaction equipment such as the sheepsfoot compacts from the bottom of the lift up, rather than from the top down. Large-tired pneumatic equipment works from the top down. In the former case, the lift thickness is limited by the length of the feet on the compactor. In the latter case, the width of the tire and the weight of the roller determine how deeply the soil will be compacted.

8.8   CHEMICAL STABILIZATION OF FILLS

There are a host of chemicals and other products that can be used to improve soil properties for a specific purpose. None is a universal cure-all for improving engineering properties or construction workability. With certain soils, certain additives have a long and widely known history of success. The principal additives employed are Portland cement and lime.

Cement is added to soil to improve strength and durability. The usual application is to pavement construction where *soil-cement* is substituted for aggregate base. Other successful uses have included slope protection for dams and levees and impermeable linings for canals and reservoirs. Cement usually proves effective as a stabilizing agent for coarse soils with a low percentage of fines or for fine soils having little plasticity. To produce true soil-cement (having high strength and adequate durability) the percentage of cement required is usually between 5 and 10 for sands and gravels and between 10 and 20 for silts and clays. Strengths approaching about one-half that of concrete are not uncommon, with values of seven-day, unconfined compressive strength in the 500 to 700 psi range generally required to meet durability requirements. Lower cement contents are sometimes used to produce *cement modified soil*, which is similar to soil-cement but does not meet durability requirements.

As for concrete, strength for soil-cement increases with time. Because development of strength begins as soon as water and cement are mixed, construction with soil-cement must be done according to a carefully controlled schedule. Long time delays between addition of water and compaction are unacceptable. Construction with soil-cement may be done by a batch process or a mix-in-place process, according to the requirements and materials for a particular job. In the batch process, constituents are obtained from stockpiled materials, mixed, transported, and placed. A very uniform product can result if close control on all phases of the work is maintained. The essential elements of a mix-in-place operation include loosening of the material to be stabilized, addition of cement, addition of water, blending, and compaction. The opportunity for nonuniformity is greater on a mix-in-place job, and sometimes a greater cement content than the actual percentage required is specified to compensate.

Lime is an effective stabilizing agent for fine soils with high plasticity. The addition of quicklime (CaO) or hydrated lime [Ca(OH)$_2$] results in both a cementing action and an alteration of the physical chemistry of the soil-water system, which both have beneficial effects. The calcium in the lime reacts with available silicas in the soil to produce a cement, which increases strength. In some cases additional silica (pozzolan) is added which enhances this effect. In highly plastic clays, the calcium from the lime is attracted to the negatively charged soil particle surfaces (see Figure 3.4) which makes their repulsive potential less strong. The result is a collapse of the double layer accompanied by decreased plasticity and improved workability of the soil. The amount of lime added does not usually exceed

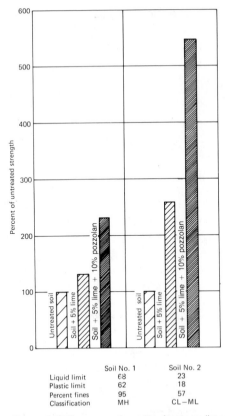

	Soil No. 1	Soil No. 2
Liquid limit	68	23
Plastic limit	62	18
Percent fines	95	57
Classification	MH	CL−ML

Figure 8.22 Summary of results of tests on lime-pozzolan stabilization. Soils compacted near plastic limit. Samples stored seven days before testing. Strength determined by unconfined compression test.

about 10 percent. Laboratory studies are used, as is also the case for soil-cement, to determine the optimum percentage. Results of one such series of tests are shown in Figure 8.22.

Hydration of quicklime by the water in a soil according to the reaction indicated by Equation 8.5

$$CaO + H_2O \rightarrow Ca(OH)_2 + Heat \tag{8.5}$$

can result in improved soil workability on a very wet construction site. Equation 8.5 also indicates that, pound for pound, quicklime has a greater potential for beneficiating soil than hydrated lime because of the proportionately greater amount of calcium (71 percent of total weight versus 54 percent) it contains. On a large job, because of this, the disadvantages of the added precautions that must be taken in handling quicklime may be offset by savings in haul costs.

Lime stabilization may be accomplished with less stringent schedule requirements than for cement stabilization, especially when the primary benefits are derived from alteration of soil-water chemistry. To this end, the lime-soil mixture is sometimes lightly compacted after blending, then remixed after a curing period, and compacted to the specified density. The curing period allows for a more thorough dispersion of calcium ions through the soil. In most instances, lime stabilization is accomplished in a mix-in-place operation.

Figure 8.23 shows a large stabilization project where cohesionless sand dredge spoils were stabilized with lime, cement, and pozzolan to form the base for a heavy industrial pavement in the backup area at a container ship terminal.

8.9 STABILIZATION OF FILL FOUNDATIONS

When fills are placed on compressible foundation soils, the weight of the fill induces settlement as discussed in Section 5.5. The purpose of the fill is usually to provide a surface for construction of structures or pavements. The amount of settlement is determined by the load resulting from the fill and the compressibility and thickness of the foundation soil. The rate at which the settlement progresses is determined by the permeability (which determines the coefficient of consolidation) of the foundation soil and the efficiency of the subsurface drainage. In the event that the settlement is large and takes a long time to occur, structures built on the fill surface may be damaged. Various methods are employed to preclude this happening.

On one project, laboratory testing and analysis showed that a planned site fill would induce some 6.5 in. of settlement at the site of a major metropolitan sewage-treatment plant. Furthermore, it was estimated that these

Figure 8.23 Stabilization of sands using lime, cement, and pozzolanic additives. Construction was by the batch process. (Photo courtesy of M. L. Byington.)

settlements would develop over about a two-year period. Since this time exceeded the planned time for construction, and since there could be no assurances that the proposed structures would not suffer damage from these deep-seated settlements, it was decided to *preload* or *surcharge* the site. In this process, a fill of greater depth than required for site development is imported with the result that greater settlements are induced at a more rapid rate than would be the case for the site fill only. The predicted relationships for this case are shown in Figure 8.24. The plan was to observe actual settlements, remove the surcharge, and proceed with construction when the settlement reached that resulting from the site fill only. The result would be a site with fill-induced settlements complete and foundation soils precompressed above the load that results from the site fill and structures. One year was allowed for the process. Actual measurements showed that the soil was much more compressible than had been anticipated, and the settlements induced were very large. Fortunately, as is usually the case, the settlements took place more rapidly than predicted. During the year allowed before removal of the surcharge and initial work on the structures, nearly 3 ft of settlement occurred, approximately half of which resulted from the site fill. The surcharge was removed and construction proceeded as planned, with no subsequent difficulty with the structures.

Surcharging accelerates settlements because of the weight of extra fill added. Settlement may also be accelerated by improving subsurface drain-

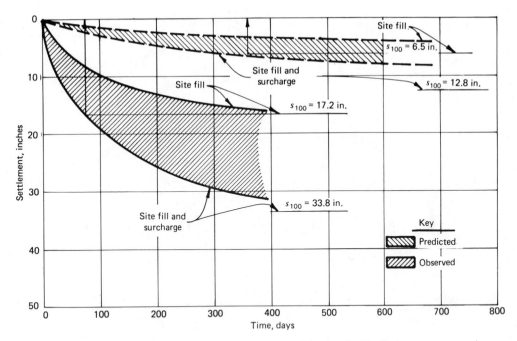

Figure 8.24 Example of time-settlement relationships for site fill effects.

age using vertical sand drains. In this process, holes are drilled in compressible, relatively impervious soils before site fill is added. The holes are relatively closely spaced, and they are filled with pervious soils. The drains so constructed greatly decrease the drainage path length in the subsoil and, as a result, speed consolidation and settlement. Figure 8.25 shows equipment used to construct over 7000 such drains to depths exceeding 150 ft on one job. This equipment consists essentially of a hollow-stem auger attached to a sand hopper that can be sealed and pressurized. The auger is drilled to the desired depth and then retracted without rotation as the sand in the hopper is forced out the auger stem to fill the resulting void. An alternative to sand drains that is currently being developed involves various fabric or plastic drains. These drains are punched into the soil with about the same spacing as sand drains and function in a similar manner. They offer a much simplified method of construction.

8.10 DEEP COMPACTION OF SANDS

If soils are to be compacted to great depths, say in lifts of up to 30 or 40 ft, special procedures are required. Coarse soils have been successfully densified by controlled explosive charges set in patterns at depth. More

Figure 8.25 Vertical sand-drain construction equipment. Sand in the hopper on the upper end of the auger is discharged under compressed air as the auger is withdrawn.

commonly, densification will be obtained by driving displacement piling, by Vibroflotation, or by use of a vibrating mandrel that is driven and withdrawn. Dynamic compaction, involving the dropping of large weights from great heights, may also be used. Displacement piling is simply timber piling that serve no purpose other than to occupy void space. If the density of a deposit is known and the desired density is established, the spacing of piling required to achieve it may be computed. A compaction piling job is illustrated

Figure 8.26 Densification of loose sand behind a bulkhead using compaction piling. (Photograph courtesy of E. G. Worth.)

in Figure 8.26. Vibroflotation is a patented process in which a vibrating probe is sunk with the aid of water jets in the deposit to be densified; the equipment is shown in Figure 8.27. As the equipment is withdrawn, the vibration continues and sand is added at the surface to occupy the volume of the depression resulting from densification. A vibrating probe such as that shown in Figure 8.28, used without the aid of water jetting, produces similar results but does not appear to be as efficient. It is, however, much less expensive. The deep compaction methods all work best in submerged clean coarse soils. Deep compaction is often specified using work-type requirements. The contractor is told what to do in terms of probe spacing, for instance. The engineer will then determine the acceptability of the work, usually with a drilling and sampling operation. If more work is required, provision should be made for additional compensation. In some instances, a performance specification is written and the contractor must know what can be accomplished with a given procedure to bid the work properly. Figure 8.29 illustrates, in a crude fashion, results obtained on several jobs of this nature. This figure may be used to estimate work requirements on a job with performance-type specifications. It is important to note, however, that the efficiency of the Vibroflotation and probing operations are time dependent and that probing time requirements may differ at different sites, even though probe spacing may be constant.

Figure 8.27 Vibroflotation equipment for densification of deep sand deposits. *(a)* Probe mounted on crane. *(b)* Probe during insertion. Note material on front loader for replacement. (Photos courtesy of Tom Huntsinger.)

Figure 8.28 Using a 30-in. pipe probe and vibratory pile hammer to densify loose sand.

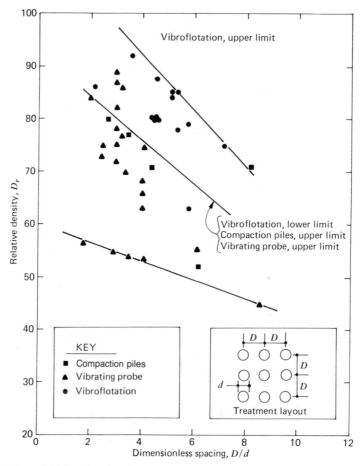

Figure 8.29 Results of various deep compaction methods for loose sands. (After Schroeder and Byington, *10th Annual Engineering Geology and Soils Engineering Symposium*, Moscow, Idaho, 1972.)

8.11   SUMMARY

Soils are compacted to produce a material with known properties, which in turn determines the suitability of an embankment. Test methods are available to serve as a standard for construction control. These same tests can be used to produce samples for laboratory studies to correlate soil engineering properties and more easily measured index properties such as density and water content.

Following study of this chapter, the reader should

1   Be able to explain why moisture-density curves for various soils exhibit the shape they do.

2 Know why soils are compacted.

3 Know in a general way how compaction influences engineering properties of soils.

4 Be familiar with common laboratory compaction testing methods.

5 Be able to perform calculations associated with compaction testing and presentation of compaction test results.

In the field, soil may be compacted in a number of ways. It should always be remembered that the purpose of compaction is to produce a manufactured fill with controlled engineering properties. The conditions under which the work is undertaken, the equipment used, and the procedures followed, all have a bearing on the final product. In many cases, the contractor is obligated by the contract language to assume the responsibility for the consequences of an unsatisfactory fill. It is, therefore, to the contractor's benefit to build it properly as well as efficiently.

The reader should now

1 Know the meanings of the terms *relative density* and *relative compaction*.

2 Be able to list the provisions of a well-written compaction specification.

3 Understand the ambiguities in a poorly written specification.

4 Know what types of compaction equipment and methods are applicable to given soil conditions.

5 Know how compaction control tests are run.

6 Understand how compaction procedures might be modified to produce a desired result.

In some contracts, provision is made for soil improvement by means other than compaction. The reader should

1 Understand the reasons for stabilization of soils by addition of lime or cement.

2 Know whether lime or cement may be the most appropriate stabilizing agent in a given situation.

3 Understand the principles involved in preloading and the use of vertical sand drains.

REFERENCES

Aldrich, H. P., "Precompression for Support of Shallow Foundations," *Journal, Soil Mechanics and Foundations Division, American Society of Civil Engineers,* **91,** No. SM2, 1965.

Anderson, R. D., "New Method for Deep Sand Vibratory Compaction," *Journal, Construction Division, American Society of Civil Engineers,* **100,** No. C01, 1974.

Asphalt Institute, *Soils Manual SM-10,* College Park, Md., 1969.

Caterpillar Tractor Company, *Caterpillar Performance Handbook* (3rd ed.), 1973.

D'Appolonia, E., *Loose Sands—Their Compaction by Vibroflotation,* American Society for Testing and Materials, Special Technical Publication No. 156, 1953.

Hall, C. E., "Compacting a Dam Foundation by Blasting," *Journal, Soil Mechanics and Foundations Division, American Society of Civil Engineers,* **88,** No. SM3, 1962.

Highway Research Board, "Cement-Treated Soil Mixtures," *Highway Research Record 36,* 1963.

Hilf, J. W., *A Rapid Method for Construction Control for Embankments of Cohesive Soil,* American Society for Testing and Materials, Special Technical Publication No. 232, 1957.

Lambe, T. W., "Soil Stabilization," in *Foundation Engineering,* G. A. Leonards, Editor, McGraw-Hill, New York, 1962.

Lee, K. L., and Singh, A., "Compaction of Granular Soils," *Proceedings, 9th Annual Engineering Geology and Soils Engineering Symposium,* Boise, Idaho, 1971.

McDonald, J. K., "Soil Classification for Compaction," *Proceedings, 10th Annual Engineering Geology and Soils Engineering Symposium,* Moscow, Idaho, 1972.

McDowell, Chester, "Stabilization of Soils with Lime, Lime-Flyash, and Other Lime Reactive Materials," *Highway Research Board Bulletin 231,* 1959.

Portland Cement Association, "Soil-Cement Construction Handbook," Chicago, 1956.

Portland Cement Association, "Soil-Cement Laboratory Handbook," Chicago, 1959.

Richart, F. E., "A Review of Theories for Sand Drains," *Transactions, American Society of Civil Engineers,* **124,** 1959.

Schroeder, W. L., and Byington, M. L., "Experiences with Compaction of Hydraulic Fills," *Proceedings, 10th Annual Engineering Geology and Soils Engineering Symposium,* Moscow, Idaho, 1972.

Schroeder, W. L., and Worth, E. G., "A Preload on Fine Silty Sand," *Proceedings of Specialty Conference on Performance of Earth and Earth-Supported Structures,* Vol. 1, American Society of Civil Engineers, 1972.

Seed, H. B., and Chan, C. K., "Structure and Strength Characteristics of Compacted Clays," *Journal, Soil Mechanics and Foundations Division, American Society of Civil Engineers,* **85,** No. SM5, 1959.

United States Bureau of Reclamation, *Earth Manual,* 1963.
Washington Department of Highways, *Standard Specifications for Road and Bridge Construction,* 1963.

PROBLEMS

1 Data points from a standard moisture-density test on a fine-grained soil are given in the following table. Reduce the data and plot the water content versus dry density curve. Clearly indicate values for optimum water content and maximum dry density from this test.

Water Content (%)	Total Density (pcf)
20.0	99.6
23.1	105.9
25.6	111.2
28.5	115.1
31.3	113.4
34.0	111.5

2 A new employee compacted one cylinder of the soil described in Problem 1 to compare new results with those shown. The data obtained included a water content of 32.5 percent and a total density of 121.9 pcf. Comment on these new findings.

3 Given the definition of relative density in terms of void ratio (Equation 8.1), derive the relationship (Equation 8.2) in terms of unit weights.

4 Derive the equation of the zero air voids curve (Equation 8.3). Plot the curves for the specific gravity of solids values 2.6, 2.7, and 2.8 on the moisture-density plot in Problem 1. Discuss the possible significance of the curves.

5 For the soil described in Problem 1, assume that a field density test on a fill lift indicates a water content and dry density of 23.2 percent and 88.6 pcf, respectively. Would specifications calling for a minimum relative compaction of 96 percent of standard maximum dry density and a water content ±2 percent of optimum be satisfied?

6 Maximum and minimum dry densities for a sandy soil according to ASTM D2049 are 96.3 pcf and 82.6 pcf, respectively. Compute the relative density of this soil if its in-place dry density were

   a. 90.2 pcf
   b. 81.6 pcf
   c. 99.1 pcf

7 Laboratory compaction test results for a plastic soil are given in the following table.

Test Method ASTM D698		Test Method ASTM D1557	
Water Content (%)	Dry Density (pcf)	Water Content (%)	Dry Density (pcf)
11.8	100.2	12.0	116.2
13.6	102.8	13.6	117.6
15.7	105.3	14.8	117.9
17.7	106.4	16.1	116.5
19.6	103.7	17.2	113.6
20.8	100.8		

Specifications call for a fill dry density of 114 pcf and further require that the fill be compacted at optimum water content for the field compactive effort. For the following field density test results on a fill, explain how you would alter the field placement operation.

a. γ_d = 115 pcf; w = 12.5 percent
b. γ_d = 112.5 pcf; w = 13.5 percent
c. γ_d = 110 pcf; w = 14.5 percent
d. γ_d = 111 pcf; w = 17.3 percent

8 A subsurface exploration report shows that the average water content of a fine-grained soil in a proposed borrow area is 22 percent and that its average natural dry density is 82 pcf. This soil is to be placed in a compacted fill with a dry density specified at 96 pcf. How many cubic yards of borrow will be required to produce 50,000 yd^3 of fill? What is the shrinkage factor for the borrow?

9 A field density test has been run on a soil containing 18 percent of its weight in particles retained on the No. 4 sieve. The dry density of the total sample was 128.7 pcf and the water content was 7 percent. The water content of the gravel fraction was 3 percent. The specific gravity of solids was 2.69. Compute the water content and dry density of the soil fraction of the fill that passes the No. 4 sieve.

10 Atterberg limits and natural water contents for five fine-grained soils are listed in the following table. Classify the soils in the Unified Soil Classification System, estimate their optimum water contents and maximum dry densities, and indicate approximately how much each soil must be wetted or dried if it is to be efficiently compacted.

Soil No.	Liquid Limit (%)	Plastic Limit (%)	Natural Water Content (%)
1	10	10	4
2	33	18	20
3	46	34	12
4	74	31	32
5	74	49	16

CHAPTER 9

DEWATERING

Many excavations are carried below groundwater level. Techniques for dealing with the problems that result are dependent on the excavation dimensions, the soil type, and the groundwater control requirements, among other factors. The simplest dewatering operations are carried out with little planning. Major operations in difficult conditions require advanced engineering and construction methods. This chapter is intended to provide the reader with

1   An awareness of the principal dewatering methods available and their applicability to various soil conditions.
2   Knowledge of some common problems associated with various dewatering methods.
3   The fundamental requirements for a dewatering system that is reliably designed.

9.1   SPECIFICATIONS

Dewatering is almost always considered the contractor's complete responsibility. He is to choose the method to be used, to be responsible for design of any appropriate system, and to ensure that it is operated correctly. In many circumstances the safety of the project, and the partially completed work, rests with the contractor's ability to conduct a successful dewatering operation. Dewatering specifications are, therefore, usually brief. The emphasis is on the assignment of the responsibility to the contractor for dewatering and protecting the work thus far completed. The sample specification that follows illustrates this point.

1B-23.1   General.   The contractor shall be responsible for the design, construction, operation, maintenance, and removal of pumping facilities, cofferdams, bulkheads, piping, etc., necessary to unwater various required work areas or divert existing water flows. Contractor shall submit plans of cofferdams and pumping facilities for care of water and for commencement of such construction. The Contracting Officer's approval of such plans shall not be construed in any way as relieving the contractor of his responsibilities for care of water, diversion, and cofferdamming, but will be primarily to ascertain compliance with the requirements of the contract (maintenance of anadromous fish run, reasonable precaution, the exercise of sound engineering judgment, and prudent construction practices), overloading or misuse of existing or new structures, the adequacy and safety of any such works, and potential damage or undermining of existing or completed works. (From *Spring Creek National Fish Hatchery Construction Contract, Skamania County, Washington,* Department of the Army, 1971.)

9.2 UNDERWATER EXCAVATIONS

In highly pervious soils or in those situations where it is not desirable to lower the groundwater table, excavations may be completed underwater. In these situations, it is necessary to first enclose the work area with some type of impervious structure. Any excavation required is completed within this structure, and the structure bottom is then sealed against the intrusion of water. When the seal has been completed, the water remaining within the structure is removed and construction is completed.

9.2.1 Caissons

The caisson method of construction involves excavation from within the permanent structure. In this method, the structure is either built in place if the site is on land or floated into position if the site is in water. When the structure or partial structure is in position, excavation from within begins. As excavation proceeds, the structure sinks because of its own weight, or weight that may be added, and the process is continued until the final foundation elevation is reached. The process is shown schematically in Figure 9.1. This method has been used successfully for small structures only a few feet in diameter and for large structures over 100 ft in width. In some instances, the top of the caisson is sealed and the work within is done under elevated air pressure. The use of compressed air permits unwatering of the caisson as excavation proceeds. As long as the air pressure within the caisson is greater than the water pressure at its base, intrusion of water is prevented.

In very soft ground, it is sometimes difficult to maintain the alignment of the caisson as the excavation proceeds. The sinking is accompanied by lurching a few inches from side to side as the structure moves through the

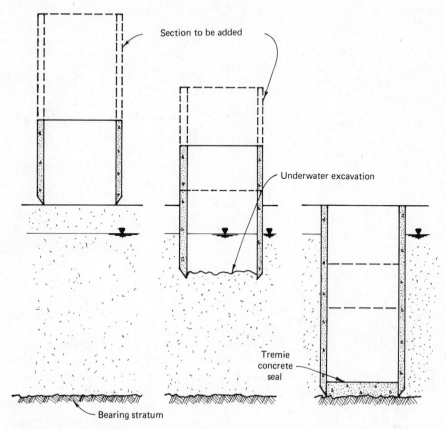

Figure 9.1 Schematic sequence of caisson installation. Part of the structure is completed before excavation begins. As excavation proceeds underwater, additional structure or weight may be added. When excavation is complete and the bearing stratum is reached, a tremie concrete seal is placed. The caisson is then unwatered and construction within completed.

ground. Large movements can overstress the structure and damage it. In other cases the frictional resistance between the caisson and the surrounding soil must be overcome by special measures to cause the caisson to sink. Adding weight is one possibility. In other cases a lubricant, usually a bentonite clay slurry, is injected through the caisson walls at the soil-structure interface. Jetting (see Section 11.4.3) may be used in cohesionless soils. The frictional resistance between the structure and the soil can be estimated in advance of construction, but such estimates are approximate. The bid price should, therefore, include the costs of measures needed to overcome these uncertainties.

When a caisson is unwatered before the seal is in place, seepage enters the bottom by upward flow from the surrounding groundwater as shown in Figure 9.2. In cohesionless soils, this upward flow can induce a *quick* condition or loss of strength at the bottom of the excavation.

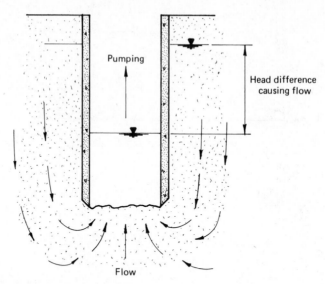

Figure 9.2 Partially unwatered caisson with no seal, showing seepage from surrounding ground.

The physical cause of the *quick* condition may be demonstrated by the analogy illustrated in Figure 9.3, in which the surface of the sand corresponds to the surface of the underwater excavation in Figure 9.2. The total vertical stress at depth L in the sand is

$$p_v = a\gamma_w + L\gamma \tag{9.1}$$

and the water pressure is

$$u = (H + a + L)\gamma_w \tag{9.2}$$

By subtracting Equation 9.2 from 9.1, we obtain the vertical effective stress

$$p_v' = L\gamma - (H + L)\gamma_w \tag{9.3}$$

Since for cohesionless soils, strength is proportional to effective stress, a condition of zero strength exists if p_v' is zero. Thus, from Equation 9.3, circumstances that may bring the zero strength condition about are those when

$$\frac{H}{L} = \left(\frac{\gamma - \gamma_w}{\gamma_w}\right) \tag{9.4}$$

That is, when the hydraulic gradient resulting from upward seepage is sufficiently large to reduce effective stress to zero. Because the total density γ of sands is approximately twice that of water, the critical value of the hydraulic gradient is about one. To prevent development of a quick condition, the head difference causing flow should be kept low. In some cases,

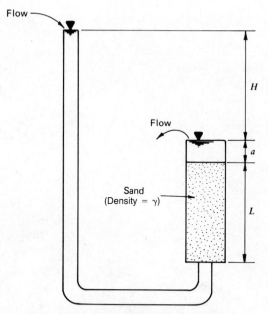

Figure 9.3 Illustration for demonstration of quick condition brought about by upward seepage in cohesionless soils.

to ensure against upward flow, the water level inside is actually maintained above the surrounding groundwater level by pumping water into the caisson.

Downward seepage along the outside of the caisson produces an increased lateral pressure on the structure. The combination of loss of support at the bottom and this increased pressure can be damaging to the structure if not accounted for in design. A further detrimental effect of this approach to dewatering a caisson is the loss of ground associated with seepage from the outside and excavation from within. As a general rule, caissons should not be used where adjacent structures exist that are susceptible to damage by loss of ground from beneath their foundations.

9.2.2 Cofferdams

Cofferdams are structures built in place to exclude water and earth from an excavation. In those instances where the distance across the excavation is sufficiently small to permit internal bracing, single-walled cofferdam construction is used. The typical application of single-wall cofferdam construction is for placement of small bridge piers. This type of operation is illustrated in Figure 9.4. In the beginning, a steel sheet pile enclosure is constructed in water. A tremie concrete seal is then poured unless piling are to be driven as shown in Figure 9.4. Driving of piling would precede the placing of the

Figure 9.4 A typical single-walled cofferdam for construction of a bridge pier foundation. Access to this cofferdam is over a temporary timber-decked trestle. Piling are being driven prior to placement of the tremie seal. Note internal struts and walers.

seal in that instance. When the seal is complete, the cofferdam is unwatered as internal bracing is installed. Construction of the pier is completed in the dry.

Cellular cofferdams are used only in those circumstances where the excavation size precludes the use of cross-excavation bracing. In that circumstance, the cofferdam must be stable by virtue of its own resistance to lateral forces. A large cellular cofferdam is shown in Figure 9.5. Individual cells within a cellular cofferdam may be of various shapes. The usual shape is the circular cell, which offers the advantage of being able to be filled as soon as it is completed. Adjacent noncircular cofferdam cells must be filled simultaneously to prevent collapse of their common walls. Unlike the single-wall cofferdam, the cellular cofferdam may be unwatered as soon as the enclosure is complete. For cofferdams on pervious foundations, this unwatering often requires the use of a supplementary dewatering system of the type discussed in later sections of this chapter. When the cofferdam has been unwatered, excavation work may then proceed in the dry.

9.3   SEEPAGE BARRIERS

For large areas where pervious soils extend to great depths, it may be impractical to construct a cofferdam only above the ground surface. In some circumstances, conditions dictate that the cofferdam extend below the ground surface to provide a barrier against seepage. No seal is constructed on the bottom of the excavation, but supplementary dewatering may be required inside the enclosure. The usual applications of the methods to be discussed include dewatering large excavations for structures and dams. The same methods are also often used for the construction of permanent seepage-control facilities.

Figure 9.5 A large cellular cofferdam for construction of Phase I, Lock and Dam 26 (Replacement), on the Mississippi River near Alton, Illinois. (Photo courtesy of the Corps of Engineers, U.S. Army.)

9.3.1 Slurry Trench Methods

In recent years, the slurry trench method has been successfully developed to deal with particularly troublesome dewatering and excavation support problems. The method involves constructing an impervious barrier beneath the ground surface. The process is illustrated schematically in Figure 9.6. As excavation for the wall is progressing, the material removed is replaced with a heavy clay slurry. The lateral pressure from the slurry is sufficient to support the excavation walls. When the excavation has been completed, concrete placement proceeds by the tremie method from the bottom to the top of the excavation. The slurry displaced by this operation is collected for reuse. When the concrete has cured, the construction site is enclosed within a rigid, impervious barrier. Alternate procedures employ excavation by dragline and backfill using clay-soil mixtures or cement-bentonite mixtures. Each procedure has advantages and disadvantages that may make it more applicable to a specific site. The work is generally performed by specialty subcontracting firms. Barriers constructed by these methods have been proven effective in nearly all foundation materials and have been constructed to depths exceeding 200 ft.

In Figure 9.7, excavation for a large project was preceded by the slurry trench-installation shown.

9.3.2 Freezing

Freezing of an impervious groundwater barrier is a means of eliminating flow to an excavation and has been used in both shallow and deep exca-

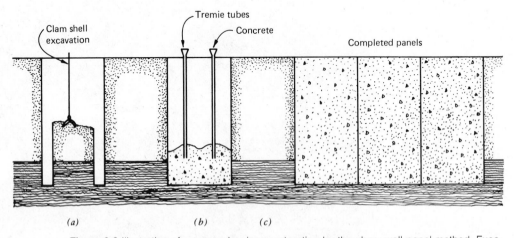

Figure 9.6 Illustration of seepage barrier construction by the slurry wall panel method. Excavation of a panel by clamshell proceeds at *(a)* with the excavation walls supported by slurry. At *(b)* a completed panel excavation is backfilled with concrete by the tremie method. The intermediate panel at *(c)* will then be completed by the same procedures as construction progresses from right to left.

Figure 9.7 Seepage barrier constructed by the slurry wall panel method for the excavation of the second powerhouse for the Bonneville Dam on the Columbia River in Washington. The wall is approximately 1 mile long and extends through ancient landslide debris containing cobbles and boulders to a depth of about 180 ft. *(a)* Exposed section with excavation beginning in background. *(b)* Close-up of exposed section, showing quality of tremie concrete and wall surface resulting from placement of concrete against trench walls.

vations. On one project, an ice barrier 8 ft thick, 130 ft deep, and about 230 ft long was formed to aid the excavation of the cutoff trench for a dam. This method was employed only after a previous attempt using upstream interceptor wells had failed.

9.3.3   Grouting

Grouting is another highly specialized, usually expensive method for retarding the flow of water. Grout curtains are used in permanent works to construct cutoffs for groundwater and sometimes have been employed as construction aids in dewatering. The process involves injection of chemical or cement grouts into the voids of pervious soils. When these grouts solidify, they form an impervious barrier. The success of the operation will depend on the distribution of the grout injected. Of course, as it is injected, the grout will follow the more pervious passages within the soil. Grouts will have varying capabilities to penetrate soils, depending on their composition. The final grout curtain position will be affected by the presence of groundwater flow. These effects are illustrated in Figure 9.8. Grouting is very much an art and does not lend itself readily to analytical treatment. The work is planned but then controlled in the field as it progresses. Surprises and increased costs are generally the rule rather than the exception. A massive grout curtain some 835 ft deep was included in the construction of the Aswan High Dam in Egypt.

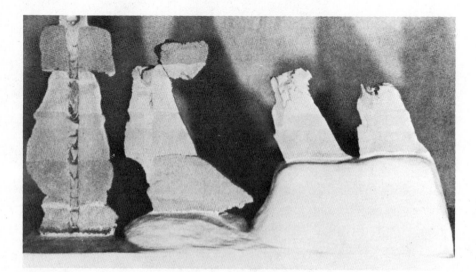

Figure 9.8 Chemical grout pattern affected by soil permeability to grout and groundwater flow. (After Karol, 1968.)

9.4   OPEN SUMPS

Open sumps are probably the most extensively used dewatering method. Sumping is simple and cheap and can be carried out with very little planning, provided that it will work at all. It is well suited to some situations and totally inapplicable to others.

9.4.1   Cohesive Soils

Sumping methods work best in soils that are nearly impervious and when an excavation must be made only a short distance below the groundwater level. The procedure is illustrated in Figure 9.9. The bottom of the work area is graded to drain to a central location where pumps are installed. In homogeneous clay soils, the excavation can usually proceed rapidly enough so that the inflow of water does not present a problem. Therefore, it is usually necessary only to install a moderate amount of pumping capacity, and the installation may be made after the excavation is complete.

Even though seepage may be readily controlled by sumps, uplift pressures may cause the base of an excavation in fine soils to heave. Heave can occur as the uplift pressure resulting from lowering the groundwater level (Equation 9.2) approaches the total pressure (Equation 9.1) in the overlying soil. Generally, well dewatering systems (Section 9.5) are required to control heave where groundwater lowering is large.

9.4.2   Cohesionless Soils

Cohesionless soils are usually of sufficiently high permeability that the success of a sumping operation will depend on the acceptability of comparatively large pumping capacities and certain problems that may arise from movement of soil particles to the sumps. In gravels and coarse sands, the

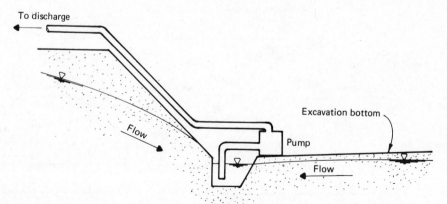

Figure 9.9 Schematic arrangement for dewatering by sumping.

principal problem to be dealt with is the need for large pumping capacity. In finer sands, the pumping capacity required is reduced, but the slopes of the excavation generally tend to ravel and run toward the sumps as the result of seepage pressures due to the groundwater lowering. This process is referred to as backward erosion, internal erosion, or piping depending on whether the particles are moving from the slope surface or from within the soil being dewatered. The effects of seepage along the slopes are discussed in Chapter 10. One unsatisfactory consequence of the movement of soil to the sump is loss of ground from the surrounding area. Cases have been recorded where the ground was actually pumped from beneath adjacent structures by this mechanism (internal erosion or piping). One means of controlling loss of ground resulting from internal erosion is to provide sumps with protective filters. A filter is a barrier to the movement of these soil particles. It may consist of a properly graded soil blanket, a metal screen, or a fabric. When sumped excavations are provided with filters, it is usually necessary to enclose the sump with a perforated conduit.

The quantity of seepage toward a sump may be estimated from the expression

$$q = \frac{k(H^2 - h^2)}{2(D - d)} \tag{9.5}$$

where q is the flow rate per unit length of sump trench and other terminology are as defined by the sketch in Figure 9.10. The usual application of Equation 9.5 would be where the amount of groundwater lowering, $H - h_t$, was known and the influence distance D was estimated. In that case, $h = h_t$ at $d = 0$ and

$$q = \frac{k(H^2 - h_t^2)}{2D} \tag{9.6}$$

The resulting discharge should be multiplied by trench length to provide the total pumping requirement.

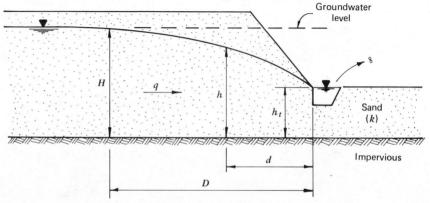

Figure 9.10 Terminology for analysis of seepage toward a sump.

9.5 WELLS AND WELL-POINT SYSTEMS

When a well is pumped, the groundwater surface in the surrounding area is depressed. The amount of depression is dependent on the pumping rate, the size of the well, the permeability of the ground, and the distance from the well. Provided that the permeability of the ground can be established with sufficient accuracy, the drawdown curve may be reliably computed. These calculations may be used to plan dewatering operations within the zone of influence of the well.

9.5.1 Single Wells

The drawdown curve for a single well penetrating an open aquifer was shown in Figure 5.6. Usually the drawdown from a single well is not of sufficient magnitude to dewater an excavation. However, single wells are often pumped in tests done prior to bidding to determine the coefficient of permeability of the ground. In these tests, all of the parameters in the governing equation for discharge are measured except the coefficient of permeability which may then be calculated. This information will be furnished to the contractor for his prebid planning. With the information so obtained, the contractor may design one of the several types of dewatering systems described in this chapter.

9.5.2 Multiple Wells

When deep wells are spaced sufficiently close so that their zones of influence overlap, the drawdown at any point is greater than the drawdown resulting from a single well. Wells of this nature may be located so that the shape of the drawdown surface fits the needs of a dewatering operation. These calculations are fairly complex. For an open aquifer such as that indicated in Figure 9.11, the expression

$$h^2 = H^2 - \sum_{i=1}^{n} \frac{q_i}{\pi k} \ln(R/r_i) \tag{9.7}$$

applies. The integer i indicates the number of any well, so that q_i is the rate at which it is pumped and r_i is the distance from the well to the point where h, the water surface height due to pumping all the wells, is to be computed. To apply Equation 9.7 for more than two or three wells involves lengthy calculations, for which a computer is appropriately employed. The drawdown surface computed for two interfering wells is shown in Figures 9.11 and 9.12, relative to the dimensions of a planned excavation.

When deep wells are used to dewater an excavation and the drawdown of the surrounding ground surface is large, several problems may arise. Loss of ground by piping through the well screens is a possibility that can

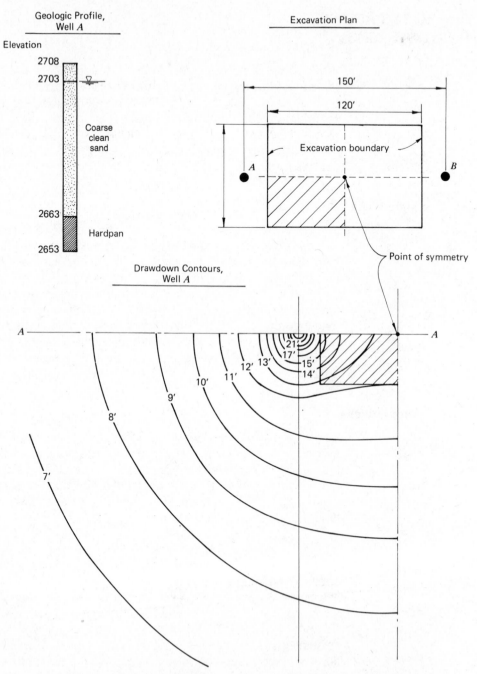

Figure 9.11 Plan view showing excavation boundary and well layout for dewatering. Drawdown water surface contours are shown for one-quarter of the excavation area.

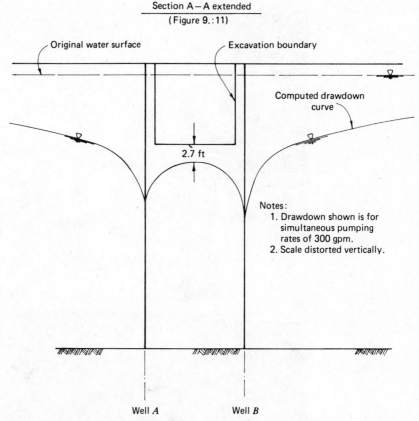

Figure 9.12 Section through excavation and wells, showing groundwater drawdown curve computed.

be overcome by proper design of the well screens to act as filters. Of potentially more serious nature is the possibility that the zone of influence of the wells may extend to such great distance that it affects the underground water supply in the surrounding area. Of further practical importance, one must consider the effects of groundwater lowering on producing compression settlements of the surrounding ground surface. Lowering of the groundwater table by 10 ft is equivalent to placing a surface load on the soil of 624 psf. The calculation of pressure changes is illustrated in Chapter 5. When the surrounding area is underlain by compressible soils, this increase in pressure produces settlements in a pattern related to the amount of groundwater lowering; an extreme example of such settlements is shown in Figure 9.13. In Figure 9.13 the groundwater lowering resulted from water supply wells and not construction dewatering. The magnitude of the settlement induced by groundwater lowering will be dependent on the compressibility of the soil and the amount of groundwater lowering. Settlement calculations were discussed in Chapter 5.

Figure 9.13 Compression settlement resulting from groundwater lowering in Mexico City. Top of well casing shown was originally at the ground surface. (Photo courtesy of J. R. Bell.)

9.5.3 Well Points

When the required groundwater lowering at a site is not large, and in some other circumstances, the use of a well-point dewatering system is advantageous. A well-point dewatering system consists of a series of closely spaced small-diameter wells installed to shallow depths. These wells are connected to a pipe or header that surrounds the excavation and that is attached to a vacuum pump. A typical well-point system is illustrated in Figure 9.14. The amount of groundwater lowering produced by well points can be readily estimated using Equation 9.6 and taking the pumping rate for each well point as the well-point spacing times q, the rate per unit length of sump. More detailed calculations are required, however, to evaluate the shape of the drawdown water surface in the vicinity of the line of well points because each well point produces a localized water surface depression. The method involves procedures for multiple wells discussed in the previous section. A quick estimate of the feasibility of using well points may be obtained by application of the nomograph in Figure 9.15. A distinct advantage of the well-point method is its adaptability in those situations where the calculations have not proven reliable. In these circumstances, the spacing of individual well points around the header can be decreased until the desired amount of lowering is obtained. There is, of course, a maximum amount of lowering that may be obtained in soils of a given permeability. Because it is a vacuum system, the amount of lowering for a single-stage well-point system is theoretically limited to a head of water corresponding to one atmosphere of pressure. Because of losses in the lines and well

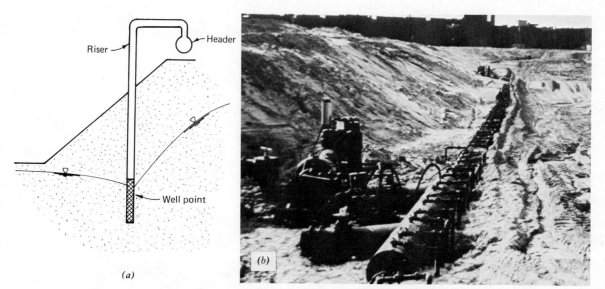

(a)

(b)

Figure 9.14 A single-stage, well-point system, showing the header and riser pipes. *(a)* Schematic, *(b)* Installation. (Photo courtesy of E. G. Worth.)

points, and other considerations, the practical limit of groundwater lowering for single-stage well-point systems is 15 to 20 ft. Where greater depths of dewatering are required, multiple-stage well-point systems may be employed or deep well systems with submersible pumps can be used.

Because well points are installed behind the toe of an excavation slope, they have the effect of creating a stable condition in the slope, since the seepage will be to the well points or into the slope. Well points may, thus, be used to stabilize a slope where slope instability has been induced by a sumping system.

Well-point equipment is specialty equipment and not usually owned by the general contractor. It may be rented or, alternatively, a specialty sub-

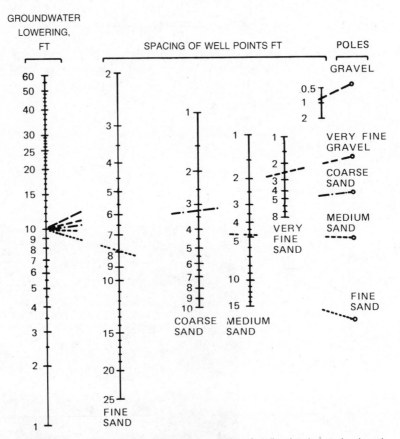

Figure 9.15 Nomograph for estimating feasibility of well-point dewatering in uniform clean sands and gravels. This nomograph should be considered as approximate only, and more detailed analyses should be completed for specific conditions at each actual project site. (Courtesy, Moretrench American Corporation.)

contractor may be employed for this particular phase of the construction contract. Several such contractors with national and worldwide experience are available. Their services include a complete range of planning, construction, and operation activities.

9.6 ELECTROOSMOSIS

Electroosmotic dewatering systems have been employed to dewater fine-grained soils. They are usually applicable to silts within a certain plasticity range, although the specific range of applicability is not clearly defined. An electroosmotic system is basically a well-point system in which the gradients causing flow are supplemented by the use of direct current electricity. A system such as this is illustrated schematically in Figure 9.16. Electrodes are inserted midway between adjacent well points. Direct current is applied so that the polarities of the well point and the electrode are as shown. The resulting electrical gradient presumably causes the positive ions in solution in the groundwater to migrate toward the negatively charged well point. The mobility of the water to the well point is thereby improved. By properly locating the well-point electrode system, the stability of an excavation slope may also be improved.

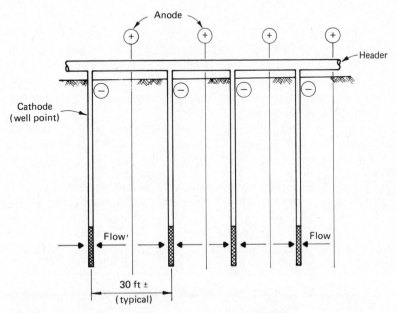

Figure 9.16 Electroosmotic dewatering system (schematic).

9.7 PLANNING DEWATERING OPERATIONS

The dewatering methods previously discussed are applicable to certain specific soil conditions and excavation sizes. Their suitability for various soil types is illustrated in Figure 9.17. Methods involving wells and well-point systems are employed where the soils to be pumped are predominantly sand and gravel. Freezing and grouting and seepage barriers may be used in these same soils. Open sumps by themselves are most applicable to clay soils; however, in the design of dewatering systems, it is the usual practice to sump excavations dewatered by other methods. Such supplementary sumps are used to control rainfall runoff within the excavation area and also to mop up that seepage which may penetrate through the principal dewatering system.

The analysis of a dewatering system requires knowledge of the permeability of the soil to be dewatered. The permeability of all but the simplest soil deposits is difficult to determine reliably. Therefore, dewatering systems are typically conservatively planned. Any analysis made for design of a dewatering system should include consideration of the effects of either underestimating or overestimating this variable. The design should then be formulated so that if the on-site conditions encountered prove to be different from those expected, the basic method of dewatering selected will still apply with some modification. In the event that deep wells are used, for instance, the pattern should permit the addition of more wells if the number originally planned is insufficient.

In practically all cases, a sudden rise in the groundwater level within a dewatered excavation cannot be tolerated. For this reason, extra pumps should be installed within the pumping system to provide standby capacity in the event of pump failure. Similarly, where electric power is used to drive pumps, generators powered by on-site sources should be available for rapid employment.

The planning of a successful dewatering operation requires the following information:

1 Ground conditions at the site to be dewatered. The essential information includes the sequence of soil strata to be encountered, their permeabilities, and the position of the natural groundwater table during the construction period.

2 The amount of groundwater lowering required.

3 The duration of the period of dewatering.

4 The applicability and costs of the various dewatering methods that may be employed.

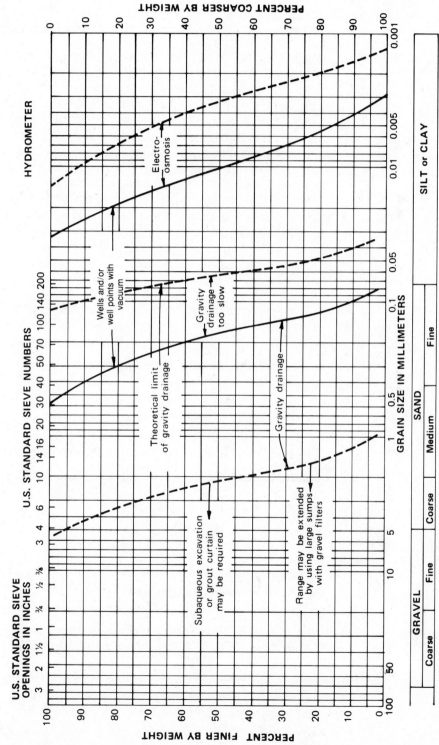

Figure 9.17 Applicability of dewatering methods. (Moretrench American Corporation.)

9.8 SUMMARY

Failure in a major dewatering effort can result in great financial loss to the contractor in that partially completed work can be destroyed. Similarly, poorly planned and executed dewatering operations can affect the performance of the completed project by altering subsurface conditions. Dewatering conducted without regard to its effects on surrounding properties can result in damages and subsequent claims against the contractor and other parties to the construction contract. Planning for dewatering should, therefore, best be undertaken by knowledgeable people within the contractor's organization. If they are not available, outside help should be sought. In either event, it must be recognized that dewatering is not a task to be undertaken lightly, nor is the planning for such an operation to be based on limited information and the hope for success. A method must be employed and a system must be designed that can reliably accomplish the work within the budget allocated. Having studied this chapter, the reader should

1 Know which types of dewatering methods are appropriate for various soil conditions.

2 Be able to anticipate problems associated with dewatering.

3 Recognize what constitutes adequate information for the design of dewatering systems.

4 Appreciate that a dewatering system, to be successful, may require considerable modification as it is built.

5 Understand that failure of the dewatering system cannot usually be tolerated and that the system should, therefore, be designed accordingly.

REFERENCES

Bowman, W. G., "Record Grout Curtain Seals Nile's Leaky Bed," *Engineering News Record,* February 29, 1968.

Casagrande, Leo, "Electro-osmotic Stabilization of Soils," *Journal, Boston Society of Civil Engineers,* January, 1952.

D'Appolonia, D. J., D'Appolonia, E., and Namy, D., "Precast and Cast in Place Diaphragm Walls Constructed Using Slurry Trench Techniques," *Proceedings of Workshop on Cut and Cover Tunneling,* Report FHWA-RD-74-57, Federal Highway Administration, Washington, D.C., 1974.

"Electro-osmosis Stabilizes Earth Dam's Tricky Foundation Clay," *Engineering News Record,* June 23, 1966.

"High Gorge Dam, Skagit River, Washington," *Engineering News Record*, January 22, 1952.

Kapp, M. S., "Slurry Trench Construction for Basement Wall of World Trade Center," *Civil Engineering*, April, 1969.

Karol, R. H., "Chemical Grouting Technology," *Journal, Soil Mechanics and Foundations Division, American Society of Civil Engineers,* **94,** No. SM1, 1968.

Mansur, C. I., and Kaufman, R. I., "Dewatering," in *Foundation Engineering,* G. A. Leonards, Editor, McGraw-Hill, New York, 1962.

Powers, J. P., *Construction Dewatering,* Wiley, New York, 1981.

Sanger, F. J., "Ground Freezing in Construction," *Journal, Soil Mechanics and Foundations Division, American Society of Civil Engineers,* **94,** No. SM1, 1968.

Terzaghi, K., and Lacroix, Y., "Mission Dam, An Earth and Rockfill Dam on a Highly Compressible Foundation," *Geotechnique,* **14,** No. 1, 1964.

Todd, D. K., *Groundwater Hydrology,* Wiley, New York, 1959.

White, R. E., "Caissons and Cofferdams," in *Foundation Engineering,* G. A. Leonards, Editor, McGraw-Hill, New York, 1962.

PROBLEMS

1 Figure 9.18 is representative of soil conditions at a construction site. Assuming that you are the contractor, select a dewatering method you would consider appropriate for each of the projects described below. Consider the information given in this and earlier chapters and any other information that may be appropriate. Describe in detail how your system would be built and operated. Justify clearly any assumptions you make. Discuss what difficulties you may encounter and how you should plan for them.

    a. A circular reservoir 100 ft in diameter is to be built with a bottom slab 25 ft below present ground surface. The groundwater level must be held at least 2 ft below the slab until the reservoir is filled.

    b. A bridge pier with a driven pile foundation is to be built with the bottom of the pile cap 40 ft below present grade. The cap is 20 ft by 30 ft in plan view.

2 Repeat Problem 1 for the following projects with soil conditions as indicated in Figure 9.19.

    a. 1200 ft of an 18-in. sewer line is to be placed in a trench about 16 ft below ground surface.

    b. A 30-ft-diameter concrete pump station is to be founded on the basalt rock at 65 ft in depth.

    c. Conditions as in *a,* except that the pipeline is 28 ft below the ground surface.

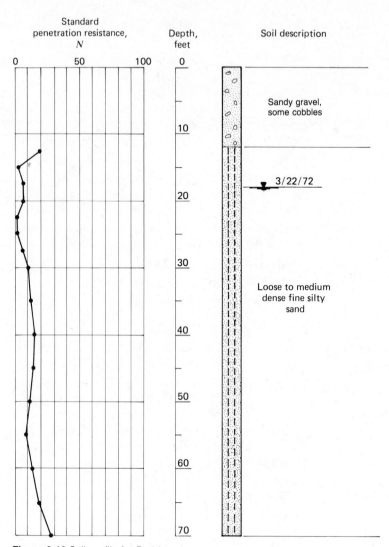

Figure 9.18 Soil profile for Problem 1.

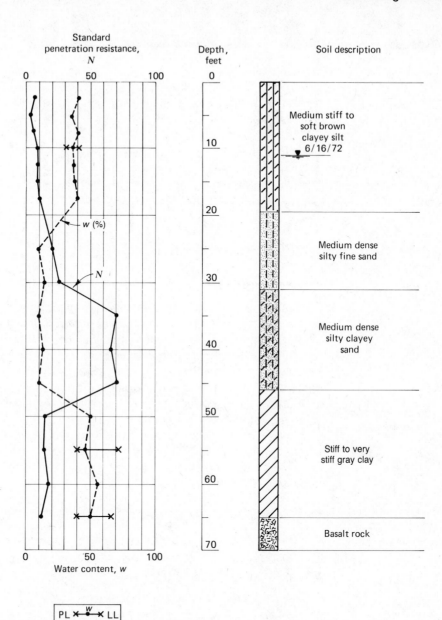

Figure 9.19 Soil profile for Problem 2.

CHAPTER 10

EXCAVATIONS AND EXCAVATION SUPPORTS

Excavation slopes would sometimes be unstable at inclinations permitted by the space available for construction. In other instances there is a need to support structures adjacent to excavations. The state of the art in excavation support technology has developed according to the needs and capabilities of the construction industry. Statutes are stringent regarding the personal safety of construction workers. For these and many other reasons, the construction contract involving significant excavation usually contains some provision covering excavation supports.

This chapter provides an introduction to various aspects of the excavation support problem. It is intended to give the reader a background bearing on the principal considerations in planning excavation support work. The topics to be covered are

1   Specification requirements in contracts.
2   Stability analysis for unsupported slopes.
3   Methods for support of shallow and deep cuts.
4   Lateral earth pressure analysis.
5   Planning for excavation supports.
6   Ground movements adjacent to braced cuts.

10.1   SPECIFICATIONS

The detail contained in specifications varies according to the complexity of the support system required for execution of the contract. When an excavation must be supported, care in the design and erection of the support system is of utmost importance. This is, of course, true because failure of excavation supports often leads to loss of life. However, for shallow excavation supports, the engineering required for successful completion of the work is not usually as comprehensive as is the case with deeper excavations. As a general rule, excavation support problems become more complex as the depth of excavation exceeds 10 to 20 ft.

Because excavation supports are not usually part of the permanent structure, the details of such systems will not be specified in the normal construction contract. The details of design and the responsibility for planning and construction are, therefore, left to the contractor. While providing workable excavation supports, the knowledgeable contractor can take advantage of past experience and do so in an economical fashion, thereby gaining a bidding advantage over competitors. On more complex jobs and when the excavation support system is to become part of the permanent structure, the specifications will be more detailed.

Requirements for excavation shoring excerpted from a construction contract where the shoring was incidental to the main work are given in the paragraph below. This is an example of the latitude that is permitted on such work.

Shoring and Sheeting Excavations and Trenches. The Contractor shall furnish and install all shoring, bracing, and sheeting required to support adjacent earth banks or structures and for the protection and safety of all personnel working in the excavations. Where sheeting and bracing are used, the widths of the excavation shall be increased accordingly. All shoring and sheeting for excavations shall remain in place until the construction has been brought up to a safe elevation and no damage will result from the removal. Trench sheeting shall remain in place until the pipe has been placed, tested for leakage, and repaired if necessary, and until the backfill at the pipe zone has been completed. All sheeting, shoring, and bracing of excavations shall conform to the requirements and recommendations of the State Industrial Accident Commission. (From *Contract Documents for the Construction of Sewerage System Additions, City of Reedsport, Oregon,* June 1969.)

On larger jobs, the method of excavation support, the disposition of these support systems, and certain performance requirements may be set forth. Excerpts from the contract for construction of a project where extensive excavation support was required are given below. In comparing this specification with the one cited previously, it is important to note that the differences lie primarily in the specification of the design and review requirements for the supports. In neither case is the responsibility for these temporary works taken from the contractor.

All excavations deeper than 6 ft shall be shored and braced, unless the slope of the cut is one horizontal on one vertical, or flatter. All excavations deeper than 30 ft shall be shored and braced regardless of the slope of the cut. The sole responsibility for the design, construction, and maintenance of all shoring rests with the Contractor. The design and installation shall be in accordance with all applicable codes and shall be subject to approval by the Contracting Officer. Review and approval by the Contracting Officer shall not in any way relieve the Contractor of his responsibilities in connection with this work.

Work of installing sheet piling, underpinning, shoring, and bracing shall be performed under direct supervision of a Specialist, experienced in supervising this kind of work. Specialists qualifications shall be submitted to the Contracting Officer for approval.

Prior to starting work, submit to the Contracting Officer for review and general comment proposed design for sheet piling, underpinning, shoring, and bracing complete with calculations, drawings and data together with sequence of operations schedule prepared by a qualified Registered Structural Engineer and bearing his seal. The Architect or the Government does not in any way assume responsibility for the design due to the above submission. The review and general comment does not imply approval. The Contractor shall be responsible for the installation of all sheet piling, underpinning, shoring and bracing and shall make good at his expense any damage caused by or due to failure of sheet piling, underpinning, shoring, and bracing, or other protection methods utilized. (From *Contract Documents for Construction of Federal Office Building,* General Services Administration, Portland, Oregon, 1972.)

10.2 EXCAVATION SAFETY

Collapse of shallow, unsupported excavations results in many construction deaths each year. Various states have adopted statutes which require that the slopes of excavations beyond some minimum depth either be supported or be inclined at an angle that will preclude their failing. These statutes generally contain provisions for enforcement of their requirements and fines to be levied against contractors who do not comply. At the federal level, the Occupational Safety and Health Act (OSHA) requires that all trenches exceeding 5 ft in depth be shored. This requirement is uniform throughout the country except where more stringent state requirements apply.

10.3 EXCAVATION SLOPES

When the construction area is large enough, excavation walls may be sloped in lieu of providing structural support. Selection of a suitable slope angle involves knowledge of the properties of the soil at the site and application of the principles of soil mechanics. The use of sloped, rather than supported, excavation walls appears to be attractive from a cost point of view. It should be remembered, however, that the savings resulting from not building a support structure are at least partially offset by the increased quantity of excavation required.

10.3.1 Slope Stability

Slopes may fail because of a number of mechanisms, depending on the nature of the soil involved and the arrangement of natural earth materials at the site in question. A number of these mechanisms are shown in Figure 10.1. Slope failures occur, of course, because forces tending to cause instability exceed those tending to resist it. Generally speaking, the driving forces are represented by a component of soil weight downslope, and the resisting forces are represented by the soil strength acting in the opposite direction. For the case of the rotational slump, driving and resisting moments about the center of slumping rather than forces are considered. In any case, the factor of safety (FS) for the slope is expressed as the ratio of the resisting forces or moments to the driving forces or moments. When the factor of safety is 1 or less, the slope must fail. When the factor of safety exceeds 1, the slope is theoretically stable. In designing cut slopes, the usual factor of safety required is between 1.3 and 1.5. In order to estimate the factor of safety for a slope, there are four required types of information. They are: (1) the soil and water profile involved, (2) the kinematics of the potential slope failure, (3) the strengths and weights of the soils, and (4) the proposed slope geometry. With this information the critical potential failure surface can be located by engineering analysis. This is a fairly simple matter in homogeneous materials. The difficulty of the problem increases with more complex subsurface conditions. The chart shown in Figure 10.2 can be used to estimate the factor of safety for cut slopes in homogeneous soils with cohesion and friction and a slope angle from the horizontal, β. For a clay soil that behaves as if $\phi = 0$, the stability number

$$N_s = \frac{\gamma H_c}{c} \tag{10.1}$$

is 3.85 for a vertical-sided trench. The values of the stability number shown in Figure 10.2 are critical values, corresponding to a factor of safety of 1. For the $\phi = 0$ case, the factor of safety of a stable slope can be calculated by finding the ratio of the critical height to the actual height or by similarly comparing the cohesion available to the critical cohesion. For instance, the stability number of a slope cut at $\beta = 65°$ in a clay for which $\phi = 0$ is 5, if the upslope angle α is zero. If the soil has $\gamma = 110$ pcf and $c = 650$ psf, and if the slope height is 20 ft,

$$H_c = \frac{cN_s}{\gamma} = \frac{650 \times 5}{110} = 29.5 \text{ ft}$$

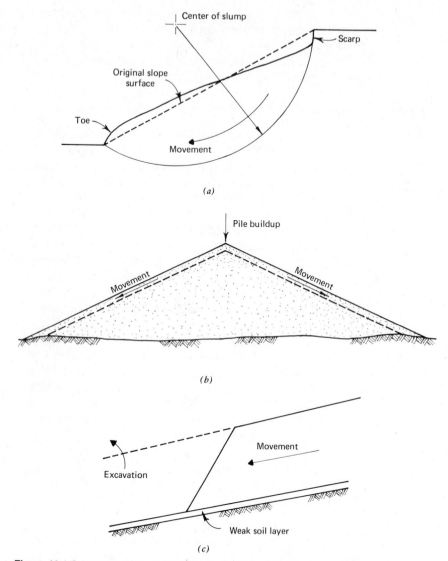

Figure 10.1 Some slope failure mechanisms. *(a)* Rotational slump in homogeneous clay *(b)* Translational slide in cohesionless sand or gravel. *(c)* Slip along plane of weakness.

Therefore,

$$FS = \frac{29.5}{20} = 1.48$$

Or, alternatively, the critical cohesion for the slope would be

$$c_c = \frac{\gamma H_c}{N_s} = \frac{110 \times 20}{5} = 440 \text{ psf}$$

And again

$$FS = \frac{650}{440} = 1.48$$

If we now assume a required factor of safety, say 1.5, and a range of consistencies as described in Table 5.3, we can tabulate theoretically safe depths for vertical cuts using the method just described. The results of such calculations are shown in Table 10.1.

Study of Table 10.1 makes it apparent that slope failures are probable in shallow excavations only for very soft to medium homogeneous clays. It can further be shown, using Figure 10.2, that significant improvement in the factor of safety for a slope of given height can be made by flattening the slope angle from 90° to about 45°.

Stability calculations for slopes in soil having both friction and cohesion can also be made using Figure 10.2, but the procedure requires several trials and is slightly more complex. We demonstrate this procedure by computing the factor of safety for a 30-ft high slope with $\phi' = 30°$, $c = 300$ psf, and $\gamma = 115$ pcf cut in level ground at $\beta = 60°$. First, we define the factor of safety as the ratio of actual soil strength on the slope to that strength required to maintain the slope at a safety factor of 1. Furthermore, we require that the factor of safety with respect to soil frictional strength be the same as that due to cohesion. Therefore,

$$FS = FS_c = FS_\phi = \frac{c}{c_{req'd}} = \frac{\tan \phi'}{\tan \phi'_{req'd}} \qquad (10.2)$$

To begin the calculation, we choose an approximate value of FS_c, say 1.5. Therefore, the strength value $c_{req'd}$ to be used in stability calculations is

$$c_{req'd} = \frac{c}{FS_c} = \frac{300}{1.5} = 200 \text{ psf}$$

TABLE 10.1 THEORETICALLY SAFE HEIGHTS FOR HOMOGENEOUS CLAY CUT SLOPES WITH VERTICAL SIDES, ESTIMATED FROM FIGURE 10.2

Soil Consistency	Unconfined Compressive Strength, q_u (psf)	Cohesion, c (psf)	Safe Height, H (ft)
Very soft	<500	<250	<5
Soft	500–1000	250–500	5–10
Medium	1000–2000	500–1000	10–20
Stiff	2000–4000	1000–2000	20–40
Very stiff	4000–8000	2000–4000	40–80
Hard	>8000	>4000	>80

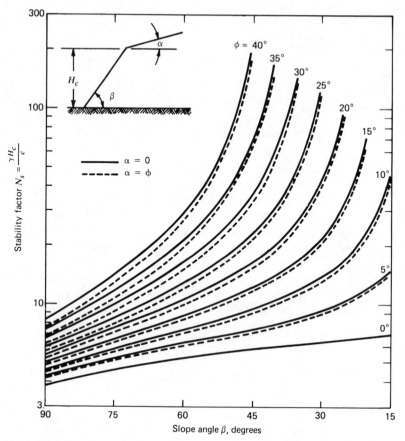

Figure 10.2 Chart for stability numbers for soils with friction and cohesion. (From *Foundation Engineering Handbook*, edited by Hans F. Winterkorn and Ysai-Yang Fang. Copyright © 1975 by Litton Educational Publishing, Inc. Reprinted by permission of Van Nostrand Reinhold Company.)

From Equation 10.2,

$$N_s = \frac{\gamma H}{c_{req'd}} = \frac{115 \times 30}{200} = 17.2$$

From Figure 10.2, for $\beta = 60°$ and $N_s = 17.2$, $\phi = 32°$. This value of ϕ is that required to maintain $FS_\phi = 1$ in the slope if the actual cohesion c is 200 pcf. The real cohesion, of course, is $c = 300$ psf, but the lower value is used in the calculations to arrive at a factor of safety. The factor of safety with respect to friction is

$$FS_\phi = \frac{\tan \phi'}{\tan_{\phi \ req'd}} = \frac{\tan 30°}{\tan 32°} = \frac{0.577}{0.625} = 0.92$$

Since $FS_\phi = 0.92 \neq FS_c = 1.5$, we must make additional trials. Assuming

$$FS_c = 1.20$$

$$c_{req'd} = \frac{300}{1.20} = 250 \text{ psf}$$

and

$$N_s = \frac{115 \times 30}{250} = 13.8$$

From Figure 10.2, $\phi_{req'd} = 28°$, and

$$FS_\phi = \frac{0.577}{0.532} = 1.08$$

Still, FS_ϕ is not equal to FS_c and an additional trial is needed. If that trial is made using $FS_c = 1.13$, FS_ϕ will be shown to also be 1.13. The theoretical safety factor for the slope is, therefore, $FS = FS_c = FS_\phi = 1.13$. This magnitude of the safety factor would indicate that, while the slope is theoretically stable, there is not sufficient soil strength to assert that it is stable enough to be as safe as normal practice would require.

By this procedure we have shown that the slope would be at critical height if its actual strength parameters were $c_{req'd}$ and $\phi'_{req'd}$, but since they are higher, according to Equation 10.2, the slope is at less than critical height and is stable at the factor of safety defined by the calculations.

If there is seepage along the slope, the factor of safety may be estimated as shown above, except that in the calculations the internal friction angle ϕ' is replaced by ϕ'_s, where

$$\phi'_s = \frac{\gamma - \gamma_w}{\gamma} \phi' \qquad (10.3)$$

It can be shown that slopes in dry or submerged cohesionless soil are stable when the slope angle is less than the soil's angle of internal friction. The factor of safety is expressed as

$$FS = \frac{\tan \phi'}{\tan \beta} \qquad (10.4)$$

The presence of seepage emerging from a slope reduces the factor of safety to half the value without seepage (Equation 10.4). A slope in partially saturated or moist sand behaves essentially as a cohesive material with friction. It may be cut, therefore, at a comparatively steep slope angle. Such slopes owe their stability to the existence of apparent cohesion. It has been previously noted (see Section 5.6.1) that apparent cohesion may be destroyed, either by drying or by saturation. Apparent cohesion, therefore, should not be relied on to maintain the overall stability of unprotected excavation slopes. A range of stable slope angles calculated for a theoretical factor of safety of 1.3 is shown in Figure 10.3. In most cases, sand slopes disturbed by excavation are at a low relative density. Low relative density results in a low angle of internal friction. Such slopes should, therefore, be designed using low assumed values of internal friction angles.

The discussion in this section has been presented to illustrate simple slope mechanics and methods of slope stability analysis. The undertaking of such analyses is in most cases considerably more complex than for the hypothetical circumstances illustrated here. Accordingly, these analyses should not be undertaken by those who do not have a thorough grounding in the principles of soil materials and slope behavior.

10.3.2 Slope Protection

In some materials, slopes that exhibit overall stability may deteriorate from the surface inward when they are exposed by construction. The result is a sloughing of soil materials into the excavation, which is troublesome and inconvenient. In these circumstances, it is sometimes advantageous to provide temporary slope protection. Generally, these types of protection are in the form of a coating or other impervious material applied to the slope.

Silty soil materials are subject to erosion during rainstorms. A number of spray-on products are available that bind the soil particles on the surface together and, at the same time, permit a permanent vegetative cover to develop on the slope. These materials may often be specified as part of the completed contract, particularly on flatter slopes. For temporary excavations, it would be unlikely that slope protection would be considered as a separate item in the contract. Where appropriate, however, the contractor should consider its use and adjust his bid items to account for the additional expense. An illustration of the use of slope protection is shown in Figure 10.4. In this excavation, the slopes consist of an almost cohesionless silty material. Direct rainfall on such slopes causes them to erode rapidly. Drying results

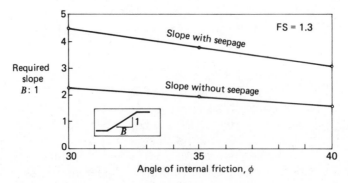

Figure 10.3 Stable slope angles in cohesionless sand.

Figure 10.4 Use of plastic sheets for slope protection in a shallow excavation.

in destruction of apparent cohesion and a similar erosion or sloughing. The plastic covering shown prevents changes in moisture content on the surface of the slope and, therefore, its stability is maintained.

In special situations where slope sloughing and raveling result in displacement of large blocks of material from the slope surface, a chain link fence has been used to provide safety in the excavation. The fencing material is simply draped over the slope surface. This application has been particularly successful where the soil tends to fail in large chunks or when the slope contains significant amounts of loose, large rock.

10.4 SUPPORT FOR SHALLOW TRENCHES

Supports for shallow trenches are not usually designed for each particular situation. Pipeline trench excavations, for instance, are usually supported by methods that have evolved and are successful in a given geographical area and in specific subsurface materials. Most of the methods employ cross-trench bracing, which somewhat restricts the work area. Several types of bracing systems for shallow cuts are illustrated in Figure 10.5. For most shallow trenches, providing the structural strength of any otherwise workable bracing system is not a problem. The most important consideration is to ensure that the bracing system is used.

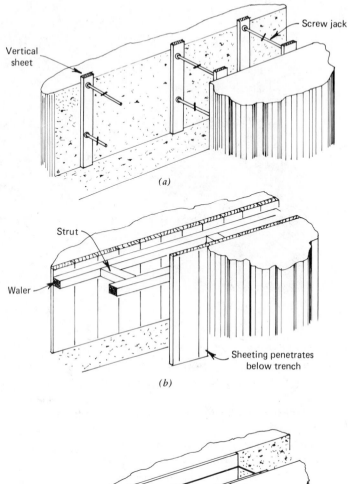

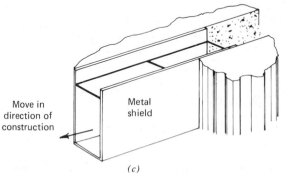

Figure 10.5 Bracing for shallow trenches. *(a)* Intermittent sheeting and bracing. *(b)* Continuous sheeting and bracing. *(c)* Trench shielding. Note that the trench shield must be so designed as to permit work within its walls. Therefore, if a pipe-laying operation is the reason for the trench, at least the rear of the shield cannot have a floor.

10.5 SUPPORT FOR DEEP CUTS

When excavation depths exceed 10 to 20 ft, generally some specialized planning for the support of such cuts is needed. Engineering for braced cuts has not progressed to the point where it is a matter of a systematic, mathematical analysis. The details of the behavior of cut bracing systems are still not yet completely understood, though much is known about them. Engineering of a deep-cut support system is very much a matter for experienced engineers and construction personnel. The design must account for the forces in the support system and the movements of the adjacent ground that will result from the excavation. Selection of a dewatering method is often a major consideration. This section discusses some of these problems.

10.5.1 Lateral Earth Pressure Calculations

Lateral pressure in an undisturbed soil deposit may be considered as proportional to the vertical pressure. The relative magnitude is dependent on the origin of the soil deposit and its stress history. To estimate the lateral pressure distribution, calculations of vertical pressure distribution are first to be made in the manner described in Section 5.2. Since water has no strength, lateral water pressure and the vertical water pressure are the same. Because soil has some strength, its tendency to spread laterally is resisted and, therefore, it generally exerts smaller lateral pressures than vertical pressures. The ratio of lateral effective pressure to vertical effective pressure at a point in a natural deposit is designated the coefficient of earth pressure at rest, K_0. For soils that have never been preloaded or overconsolidated, this coefficient varies between about 0.3 and 0.5 and is usually calculated as

$$K_0 = (1 - \sin \phi') \tag{10.5}$$

Preloading or overconsolidation, discussed in Chapter 5, refers to those situations where an existing soil deposit has been subjected to greater vertical load than presently exists. This situation generally arises where the surface of the deposit has undergone major erosion or where it has carried the added weight of ice during glacial periods. It is also possible to create the effects of overconsolidation by drying in a severely arid climate. In these situations, the lateral pressure exerted within a soil deposit may exceed the vertical pressures.

Excavation supports are associated with cuts. As a cut is made, the soil at the face tends to expand or move into the cut area. This expansion mobilizes some of the soil's strength. If a support is placed against the excavation surface in such a manner as to prevent all movement, then the stress existing before excavation is maintained. If some movement is permitted, the distribution of pressure changes. The actual distribution of pressure will depend on the mode of movement permitted. When a retaining structure is permitted to rotate a small amount so that the top of the structure moves out with respect to its base, the distribution of pressure remains linear and increasing with depth, as shown in Figure 10.6a. It is reduced to a limiting value, proportional to depth, where the constant of proportionality is K_A, the active lateral earth pressure coefficient. This may be computed from theoretical considerations. It is generally not greatly different from the coefficient of earth pressure at rest, although it is smaller.

Excavation supports tend to move outward at the base with respect to the top, depending on their method of construction. In these situations, the soil pressure at the bottom is reduced and the soil pressure at the top increases. The total force against the support is equal to or slightly greater than the total force for the active case; however, it acts higher on the structure. This is an important observation to consider in designing excavation supports. Pressure distributions similar to those shown in Figure 10.6b cannot be computed from theoretical considerations. Generally, we rely on empirical methods derived by observing forces on actual excavation support systems. Engineers at many locations throughout the world have measured loads in the struts supporting various types of sheeting and bracing systems. From these strut loads, back calculations of apparent earth pressures acting on the sheeting have been computed. Envelopes of maximum values of these apparent pressures have been constructed. These are shown in Figure 10.7. It is important to note that any water pressure existing must be added to these apparent earth pressures to arrive at the total lateral pressure on a cut support. It is further important to note that these envelopes represent

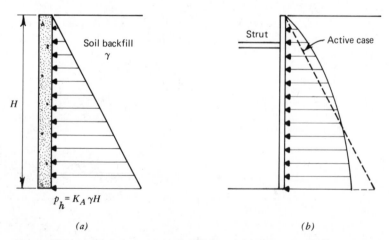

$$p_h = K_A \gamma H$$

(a)   (b)

Figure 10.6 Lateral earth pressure distribution. (a) Active case. (b) Pressure on excavation bracing.

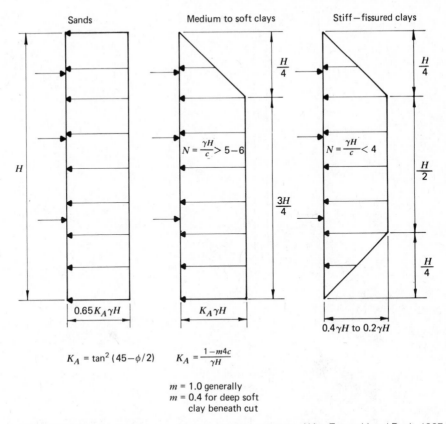

Figure 10.7 Apparent earth pressure on excavation supports. (After Terzaghi and Peck, 1967.)

fictitious pressures that, when applied to an excavation support for design, will permit estimates of the maximum values of strut loads at any level. Measured individual strut loads would be expected to differ and generally would be lower than those calculated. This produces an inherent conservatism in excavation support design that is warranted by the catastrophic consequences of a cut support failure. Failure of a single strut in a long excavation results in the transfer of the load it carried to adjacent members. If these members fail, their load is in turn transferred to adjacent members and, with each load transfer, the amount of load to be carried increases. Thus, if a single strut fails, it is quite likely that the failure will lead to a collapse of the entire system.

10.5.2 Excavation Support Methods

Excavations may be supported by either cantilever structures or structures with some form of additional support. The two possibilities are schematically diagrammed in Figure 10.8. The supported system is in most common use.

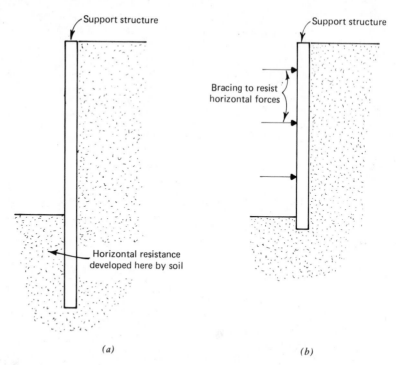

Figure 10.8 General methods of excavation support. *(a)* Cantilever systems. *(b)* Supported systems.

It consists of two essential elements: that part of the structure that is in direct contact with the soil, and that might be referred to as the *soil support*, and the *bracing* structure that supports it. In the cantilever system, a single structure serves both of these functions. The resisting capability of a cantilever structure is dependent on its rigidity and the support provided by the foundation soil below the base of the excavation. Cantilever structures are usually restricted to use in shallow excavations. The range of depth may be extended by using a rigid wall with a great embedment depth. An example of this approach is the *cylinder pile* method in which large-diameter concrete cylinders are cast adjacent to one another in augered holes to form a cantilever wall. This method has been used to successfully support excavations to 50 ft deep and more.

A number of methods for providing soil support have been developed. One of the earliest is the method employing driven *sheeting*. In this method, either timber, concrete, or steel sheets are driven ahead of the excavation. These sheets are then braced from the inside as the excavation proceeds. A more popular method that has become highly developed is the use of the *soldier pile* or the *soldier beam and lagging* system. In this method, H-piling are set in predrilled holes around the periphery of the excavation. Predrilling as opposed to driving is used to provide close control of alignment and location. These piles are then grouted in place with weak concrete. As the excavation begins, the soil and concrete are carefully trimmed away from the soldier pile and the horizontal lagging is installed between adjacent pile flanges. In some cases, the lagging is eliminated when the soils are highly cohesive and are capable of arching between soldier piles. Even in these instances, however, a surface binder such as gunnite is often applied to the soil to prevent raveling. The soldier pile and lagging method obviously is inappropriate for perfectly cohesionless soil. In this type of soil, a sheeting method must be used.

For very difficult conditions in soft ground with a high water table, the *soldier pile-tremie concrete* method has been developed. In this method, soldier piles are set in predrilled holes, and the entire space between flanges of adjacent soldier piles is excavated in a slurry trench as shown in Figure 10.9. Very specialized equipment is used for this purpose. When that excavation has been completed, tremie concrete is placed and reinforcement is set, if appropriate. The result is a continuous concrete wall cast beneath the surface of the ground prior to the beginning of any excavation. When this wall is intact and complete, excavation and interior bracing can begin.

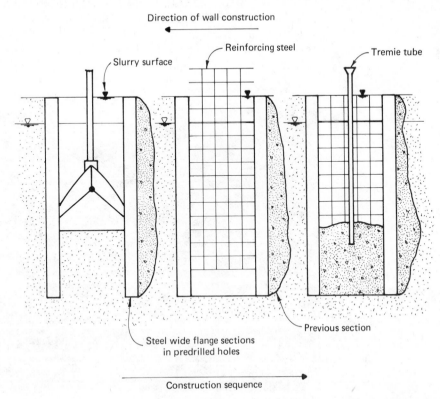

Figure 10.9 Construction of underground concrete wall by soldier pile-tremie concrete method, including excavation, placement of reinforcing steel, and tremie concreting.

When soil support is in place, the construction of structural support or bracing begins. For narrow excavations, internal struts are often most appropriate. When the excavation level reaches the point where struts are to be installed, a horizontal member called a *waler* is placed against the soil support. Intermediate struts are then installed from waler to waler across the excavation. A typical *cross-lot strutting* system is shown in Figure 10.10. It is obvious that the presence of the struts restricts the working space within the excavation.

For very wide excavations, it is often not feasible or reasonable to use cross-lot struts. In these circumstances, *raker bracing* such as that shown in Figure 10.11 can be used. To install a raker bracing system, the central portion of the excavation is completed by sloping the excavation boundaries. When the required depth is reached, support for the bottom of the rakers can be installed. These supports will be either driven piling or footings. Construction of the soil support and removal of the remainder of the excavation then begins. Raker bracing is installed as the room becomes available. A raker bracing system offers some advantage over cross-lot bracing in that the central portion of the work area is comparatively uncluttered.

The use of *tiebacks* is a very popular alternative to the installation of internal bracing. To install tiebacks, inclined holes are drilled through the soil support system as the excavation proceeds. A completed project is shown in Figure 10.12. Figure 10.13 shows a tied-back slurry wall constructed to support a

Figure 10.10 A strutted excavation.

Figure 10.11 Raker bracing for soldier pile and lagging excavation supports.

deep cut and obviate the need for dewatering outside the cut area. The tieback installed in each hole may be a concrete prestressing tendon or simply a steel rod. In either case, it is grouted within the stratum that is to provide the support and then tensioned using hydraulic jacks pushing against the soil support system. It is general practice to pretest all tieback anchors to loads greater than the anticipated design load. The load in the anchor is then set somewhere near the anticipated design load. Because of this pretensioning, tieback systems are generally very successful in preventing movements of the excavation walls. Estimates of anchor capacity can be made using methods for estimating the capacity of foundation piling, but are usually considered as estimates only, because installation methods affect anchor size and soil disturbance. Final design is based on pretest results.

When permanent construction inside a braced excavation is complete, it is the usual practice to leave the wall in place. Often, the wall is used as the back form for the permanent basement of the structure. It may be necessary, however, in some instances, to remove all of the bracing wall. Examples would be when it encroaches on a public right of way or on adjacent private property that might be used for subsequent subsurface construction. Tiebacks, if left in place, are always cut to relieve tension when the permanent structure can safely carry the load. In some cases, particularly with heavy stiff walls constructed using tremie methods, the temporary walls become part of the permanent construction.

Figure 10.12 A tied-back excavation. (Photo courtesy of George B. LaBaun.)

10.6 PLANNING FOR EXCAVATION SUPPORTS

An outline for planning an excavation support system that has been suggested is given in Table 10.2. This outline illustrates the fact that the planning effort will encompass activities that are not necessarily part of the contractor's operation. In some cases, all of the activities may be his responsibility. In other cases, he may be asked to build an excavation support system that has been completely specified. In either case, however, the steps in planning could be essentially the same. Only the responsible parties may change. It is of primary importance to establish as accurately as possible the nature of soil and subsurface conditions. Groundwater conditions are a primary consideration. The condition of adjacent structures and utilities must also be established before excavation begins. This is to aid in planning the excavation work and the support system to preclude damage to these facilities. It also provides a very important baseline of information against which claims for damage resulting from construction may be compared. For this reason a requirement for a preconstruction survey may be specified, as indicated below.

Prior to excavation, the Contractor and Construction Engineer shall jointly survey the condition of the adjoining properties. The Contractor shall take photographs and make records of any prior settlement or cracking of structures, pavements,

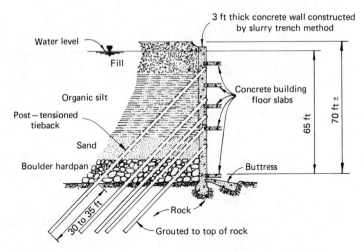

Figure 10.13 Tied-back concrete wall constructed underground for excavation support and dewatering at New York's World Trade Center. (After Kapp, 1969.)

TABLE 10.2   ENGINEERING AN EXCAVATION

Step No.	Activity	Considerations
1	Explore and test subsoil.	
2	Select dimensions of excavation.	Structure size and grade requirements, depth to good soil, depth to float structure, stability requirements.
3	Survey adjacent structures and utilities.	Size, type, age, location, condition.
4	Establish permissible movements.	
5	Select bracing (if needed) and construction scheme.	Local experience, cost, time available, depth of wall, type of wall, type and spacing of braces, dewater excavation sequences, prestress.
6	Predict movements caused by excavation and dewatering.	
7	Compare predicted with permissible movements.	
8	Alter bracing and construction scheme, if needed.	
9	Instrument-monitor construction and alter bracing and construction as needed.	

Source: T. W. Lambe, ''Braced Excavations,'' *Proceedings, Specialty Conference on Lateral Stresses in the Ground and Design of Earth Retaining Structures,* American Society of Civil Engineers, 1970.

and the like, that may become the subject of possible damage claims. Such recorded damage shall be suitably identified on the structure or pavement, and an official list of damage shall be prepared and signed by all parties making the survey to indicate their concurrence with survey findings. A copy of damage records shall be submitted to the Contracting Officer. (From *Contract Documents for Construction of Federal Office Building,* General Services Administration, Portland, Oregon, 1972.)

A series of trials leads to the final establishment of a bracing and construction scheme for excavation supports. Experience, such as that documented in Figure 10.14 concerning movements, is an invaluable aid in the process. Figure 10.14 was developed from data gathered on a number of braced excavation projects. It represents the ability of contractors to control ground movements for widely varying subsurface conditions, and incorporates the fact that these abilities will vary from job to job. As such, it is an empirical, rather than a theoretical, expression of actual expectations. To estimate settlement induced adjacent to an excavation, one would first choose the point on the abscissa that represents the location at which the

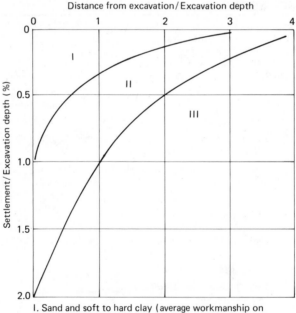

I. Sand and soft to hard clay (average workmanship on support construction).

II. Very soft to soft clay — Condition I (construction difficulties).

III. Very soft to soft clay — Significant depth below excavation bottom.

Figure 10.14 Movement limits associated with braced excavation supports. (After Peck, 1969.)

estimated settlement is desired. Say we have a 30-ft-deep excavation in soft clay, and the settlement 15 ft from the bracing wall must be estimated. At the point 0.5 on the abscissa, and for the line dividing regions I and II on the plot, we read, on the ordinate, about 0.6. This would indicate that the expected settlement would be about 0.006 × 30 or 0.18 ft. It has also been shown that horizontal movements toward the excavation of approximately equal magnitude should be planned for.

Aside from subsurface conditions and excavation dimensions, other factors, including dewatering, sequence of bracing installation, construction of the support wall, underpinning, prestressing of struts, and foundation construction, influence the magnitude of deformations. For this reason, Figure 10.14 cannot be used reliably for every job. It represents only a guide and a starting point for more detailed engineering planning. Theoretical methods are available and can be very helpful in estimating deformations caused by excavations. The methods cannot, however, simulate all of the construction activities involved. They are, therefore, best used as a supplement to informed judgment.

Like dewatering systems excavation support systems are often redesigned during construction to accommodate unanticipated conditions. For that reason, on major excavation work, support systems are frequently monitored by measuring deflections and forces in key members. If the performance of the support system is not within the requirements of the work, it is then necessary to alter the design. Performance observations serve this very important function. They also permit us to better plan future excavations by providing the data base needed to quantify our experiences.

10.7 SUMMARY

This section has discussed matters of considerable importance related to the stability of slopes for excavations and to methods of preventing the movements of excavation walls. The discussion given has been incomplete in the engineering sense, but has provided a look and some of the basics involved in these operations. It must be remembered that the planning of major excavation work should be left to those with extensive experience in these matters, for the consequences of failure can be heavy in terms of loss of life and property damage.

Following study of this chapter and solving of the suggested problems, the reader should

1 Appreciate the responsibilities of the contractor in projects where excavation supports are required.

2 Understand the fundamental concepts of simple slope stability analysis.

3   Be able to calculate lateral pressures in a soil deposit.

4   Know the requirements for planning each of the steps involved in constructing excavation supports.

5   Be able to describe basic methods for supporting walls of open excavations.

REFERENCES

Darragh, R. D., "Tiebacks, Type of Installation—Analysis," *Seminar on Design and Construction of Deep Retained Excavations,* American Society of Civil Engineers, Structural Engineers Association of Northern California, 1970.

Fang, H. Y., "Stability of Earth Slopes," in *Foundation Engineering Handbook,* Hans F. Winterkorn and H. Y. Fang, Editors, Van Nostrand Reinhold, New York, 1975.

Gerwick, B. C., "Slurry Trench Techniques for Diaphragm Walls in Deep Foundation Construction," *Civil Engineering,* December 1967.

Gould, J. P., "Lateral Pressures on Rigid Permanent Structures," *Proceedings, Specialty Conference on Lateral Stresses in the Ground and Design of Earth-Retaining Structures,* American Society of Civil Engineers, 1970.

Lambe, T. W., "Braced Excavations," *Proceedings, Specialty Conference on Lateral Stresses in the Ground and Design of Earth-Retaining Structures,* American Society of Civil Engineers, 1970.

O'Rourke, T. D., "Ground Movements Caused by Braced Excavations," *Journal, Geotechnical Engineering Division,* American Society of Civil Engineers, 107, No. GT9, 1981.

Peck, R. B., "Deep Excavations and Tunneling in Soft Ground," *Proceedings, 7th International Conference on Soil Mechanics and Foundation Engineering,* State of the Art Volume, 1969.

Taylor, D. W., *Fundamentals of Soil Mechanics,* New York, Wiley, 1948.

Terzaghi, K., and Peck, R. B., *Soil Mechanics in Engineering Practice,* New York, Wiley, 1967.

PROBLEMS

1   A 20-ft-deep cut for a basement excavation is specified to be sloped at 1:2 (one horizontal to two vertical). The upper 40 ft of soil at the site consists of a homogeneous clay with a density of 90 pcf and an unconfined compressive strength averaging 2400 psf. The water table is at great depth. Can the cut be made? If so, what is the factor of safety for

the slope? Is the slope safe enough for normal work to progress in the excavation without danger?

2 A vertical sided trench was being cut in a very soft clay soil with a density of about 110 pcf. A depth of 8 ft had been reached when the sides of the trench caved in. Could the trench be completed to a depth of 10 ft by sloping the sides before installing bracing? If so, how steeply may the sides be cut? Show your calculations and sketch the procedure you would follow to complete the job.

3 Assuming the soils at the site described in Figure 5.16 are normally consolidated, estimate the maximum values of the lateral total, water, and effective pressures in the natural ground. Present your results in a sketch showing variations with depth.

4 A 25-ft-diameter concrete caisson is to be sunk to the rock surface shown in Figure 5.16 at a depth of 50 ft. Describe, using sketches where appropriate, how you would do it, and what problems you might encounter.

5 A large (200 ft square) structure is to be built on a downtown site where subsurface conditions are as shown in Figure 10.15. The structure foundations are to be placed on the surface of the dense sand. A 4-ft-deep gravel fill is then to be placed above the sand to support the basement

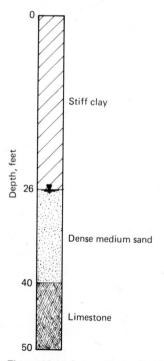

Figure 10.15 Soil profile for Problem 5.

floor. Streets and sidewalks surround the property. The structure walls proposed extend to within 5 ft of the sidewalk line. Propose a method for supporting the excavation walls. Use sketches and justify your assumptions.

6 Repeat Problem 5, assuming that the structure is to be founded on the rock.

CHAPTER 11

FOUNDATION CONSTRUCTION

Most structures are built on shallow foundations consisting of footings or rafts. The latter are simply large footings that cover the building's entire base. In a structure that has been designed by an engineer, the foundation sizes have been selected to control settlement and provide an adequate margin of safety against shear failure in the foundation soils resulting from building loads. Careful attention to the specified details of construction is necessary to ensure that foundation performance is in accord with design assumptions. In this chapter it will be shown

1 How settlement estimates and bearing-capacity calculations are made.
2 Why the usual provisions of specifications for shallow foundation construction are important.

Deep foundations (piles and piers) are used to bypass unsatisfactory soils and transfer structure loads to a suitable bearing stratum. Pile construction is fraught with uncertainty and because of this has become associated with a certain amount of mystery. Some of the mystery is imagined and some is very real. Controversy surrounds many aspects of pile construction, probably more so than any other type of work. This chapter will consider the fundamentals of the subject in an attempt to dispel some of the misconceptions that seem to be part of many such controversies. It will also deal with the construction of deep foundations using drilled piers. The reader will learn

1 Why piling and piers are used and how they carry loads.
2 Basic pile types and materials and the advantages and disadvantages of each.
3 The principal types of pile-driving and pier-drilling equipment and their applicability to certain job conditions.

4 Methods of determining deep foundation load-carrying capacity.

5 The usual provisions in a pile-driving contract.

6 About some common problems arising in pile and pier construction, and how they are dealt with.

11.1 SETTLEMENT ANALYSIS

Settlement forecasts are always a part of foundation design analysis. Understanding how these analyses are made contributes to an understanding of certain provisions in construction contracts. It may also be useful to know the basics of settlement calculations in those instances where temporary foundations must be installed.

11.1.1 Footings on Cohesive Soils

Equation 5.21 provides the basis for consolidation settlement calculations on cohesive soils. Consolidation settlements result from a decrease in the soil's void ratio under load, and they occur at a rate governed by soil permeability and drainage. On all soils, immediate settlements occur as a result of elastic distortion of the area influenced by the footing load. These settlements are usually quite small on overconsolidated cohesive materials. To use Equation 5.21 the compression index of the soil must be known, along with the effective pressures in the soil before and after loading. Figure 11.1 shows a hypothetical problem that we can use to demonstrate the essentials of a settlement analysis.

Assume the square footing in Figure 11.1 rests on the surface of an overconsolidated clay soil for which tests have shown that γ = 110 pcf, e = 0.76, p'_c is 10,000 psf, and the slope of the recompression curve (see Figure 5.9) is 0.043. The first step in the settlement analysis is to determine the initial stresses and final stress in the soil profile for the depth of significant influence beneath the footing. The calculations are made for a number of layers of soil of increasing size. For our illustration we have chosen layers designated 1 to 4. The pressure calculations are made at the middepth of each layer. The initial vertical effective stresses are

Level a $\quad p'_{1a}$ = 110 × 2 = 220 psf

Level b $\quad p'_{1b}$ = (110 × 6) − (62.4 × 2) = 535 psf

Level c $\quad p'_{1c}$ = (110 × 12) − (62.4 × 8) = 821 psf

Level d $\quad p'_{1d}$ = (110 × 20) − (62.4 × 16) = 1202 psf

The contact pressure at the footing base is the column load divided by the footing area. The pressure resulting from the column load decreases with

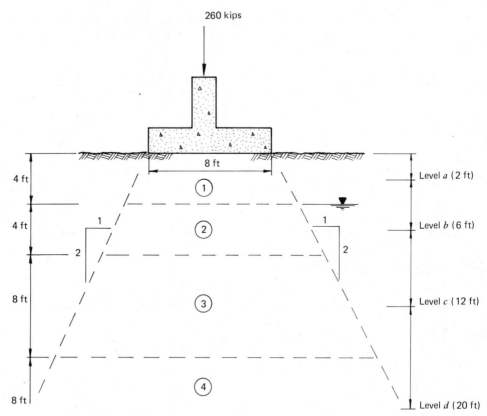

260 kips

8 ft

4 ft

4 ft

8 ft

8 ft

Level a (2 ft)

Level b (6 ft)

Level c (12 ft)

Level d (20 ft)

Figure 11.1 Problem for settlement analysis illustration.

depth beneath the footing. Reasonable estimates of the pressure at any depth may be made by dividing the column load by an area, derived by increasing the dimensions of the area influenced by the depth in the soil beneath footing grade. This is commonly known as the 2:1 rule. Application to the footing shown yields

Level a $\Delta p_a' = 260{,}000 \div 10^2 = 2600$ psf

Level b $\Delta p_b' = 260{,}000 \div 14^2 = 1327$ psf

Level c $\Delta p_c' = 260{,}000 \div 20^2 = 650$ psf

Level d $\Delta p_d' = 260{,}000 \div 28^2 = 332$ psf

Adding the stress increment to the initial stresses,

Level a $p_{2a}' = 220 + 2600 = 2820$ psf

Level b $p_{2b}' = 535 + 1327 = 1862$ psf

Level c $p_{2c}' = 821 + 650 = 1471$ psf

Level d $p_{2d}' = 1202 + 332 = 1534$ psf

The second step is to make the settlement calculations for each layer using Equation 5.21, and sum those results to obtain the settlement estimate. The results are shown in Table 11.1.

TABLE 11.1   RESULTS OF SETTLEMENT ANALYSIS ILLUSTRATION

Soil Layer	h_1 Layer Thickness (in.)	$\dfrac{C_c}{1 + e_1}$	p_1' (psf)	p_2' (psf)	Δh (in.)
1	48	0.0244	220	2820	1.3
2	48	0.0244	535	1862	0.6
3	96	0.0244	821	1471	0.6
4	96	0.0244	1202	1534	0.2
			Total estimated settlement		2.7

11.1.2 Footings on Cohesionless Soils

Settlements of footings on cohesionless soils generally occur as soon as the load is applied and result from elastic distortion of the ground without significant change in the soil's void ratio. Because of this, and the difficulty in securing and testing samples, Equation 5.21 is usually not used for settlement calculations. Instead, these estimates are made from empirical correlations among settlement, bearing pressure, standard penetration resistance, and footing size derived from load tests, settlement observations, and field testing. If N is the average standard penetration resistance within a depth below the footing equal to its width,

$$s = \frac{8q}{N} \qquad (B < 4 \text{ ft}) \qquad (11.1)$$

$$s = \frac{12q}{N}\left(\frac{B}{B + 1}\right)^2 \qquad (B > 4 \text{ ft}) \qquad (11.2)$$

In these expressions B is the footing width in feet, q is the bearing pressure beneath the footing in tons per square foot, and s is the estimated settlement in inches. For design purposes, the settlements calculated from Equations 11.1 and 11.2 are normally doubled if the foundation soils are submerged. Observations have shown that the use of these equations results in very conservative (smaller than actual) settlement predictions, and that actual settlements are about one-half of the calculated values, without the correction indicated above for the presence of groundwater beneath the footing. Thus, if the footing in Figure 11.1 was on sand for which $N = 18$ instead of the clay indicated, design settlement analysis would indicate

$$s = \frac{12 \times 260,000}{18 \times 64} \times \frac{1}{2000}\left(\frac{8}{9}\right)^2 \times 2 = 2.1 \text{ in.}$$

Whereas, in reality, the actual settlement would probably be on the order of

$$s = \frac{12 \times 260{,}000}{18 \times 64} \times \frac{1}{2000}\left(\frac{8}{9}\right)^2 \times \frac{1}{2} = 0.5 \text{ in.}$$

It is possible to obtain more reliable settlement estimates for foundations on sands than is indicated by the above procedures. The methods for doing so, however, involve extensive laboratory testing and analysis and are not warranted for design purposes in all but very special cases.

11.2 FOUNDATION BEARING FAILURE

While foundations may settle excessively under working load, there is also the possibility that they may be so heavily loaded that the strength of the supporting ground is inadequate to resist a punching type of sudden failure. An example is shown in Figure 11.2 where the mudsills supporting the form shoring on a bridge project failed during a concrete pour. Similar *bearing failures* often occur beneath fills on soft ground. Design analyses for these situations are directed to providing an adequate factor of safety against bearing failures. The fundamentals of the methods employed are outlined here.

11.2.1 Shallow Foundations

Figure 11.3 illustrates a generalized situation in which an area of width B is loaded by a pressure of intensity q at depth D_f in the ground. The pressure is presumed to result from a foundation load.

If the pressure intensity is increased to the point where a ground failure is produced, it will have a value q_d that may be estimated from

$$q_d = \frac{\gamma B}{2} N_\gamma + \gamma D_f N_q + c N_c \tag{11.3}$$

if the loaded area is very long and

$$q_d = \frac{\gamma B}{2} N_\gamma + \gamma D_f N_q + 1.2 c N_c \tag{11.4}$$

if the loaded area is square. The intensity q_d is the ultimate bearing capacity, γ is the soil density, and c is its cohesive strength below a-a. The terms N_γ, N_q, and N_c are factors derived theoretically that are functions of the soil's internal friction angle ϕ.

For square footings on clay for which $\phi = 0$

$$q_d = \gamma D_f + 6.2c \tag{11.5}$$

Figure 11.2 Failure of shoring foundations during a concrete pour.

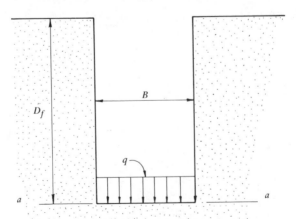

Figure 11.3 Terminology for bearing-capacity analysis.

and for footings on sands with $c = 0$

$$q_d = \frac{\gamma B}{2} N_\gamma + \gamma D_f N_q \qquad (11.6)$$

The factors N_γ and N_q may be estimated on the basis of standard penetration test results or laboratory tests from Figure 11.4.

The allowable pressure on the foundation q_a is selected by dividing q_d by a factor of safety selected with consideration for the reliability of the information available for analysis, including soil properties and probable loads. The value of the safety factor is usually about 3.

To illustrate the application of these methods, consider the footing shown on Figure 11.1. If, for the case of a cohesive soil, the strength c is 3000 psf, then from Equation 11.5

$$q_d = 6.2 \times 3000 = 18,600 \text{ psf}$$

and, since the actual bearing pressure under the load shown is 4062 psf, the factor of safety against bearing failure is

$$FS = \frac{q_d}{q_a} = \frac{18,600}{4062} = 4.58$$

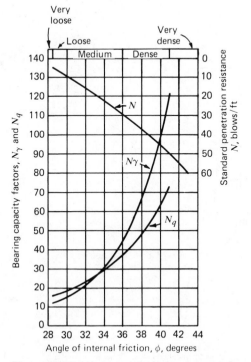

Figure 11.4 Bearing-capacity factors for cohesionless soils. (From R. B. Peck, W. E. Hanson, and T. H. Thornburn, *Foundation Engineering*, Wiley, New York, 1974.)

which would normally be considered adequate. If the footing were on sand with $\gamma = 110$ pcf and $N = 18$ as previously assumed for the settlement analysis illustration in Section 11.1.2, then, from Equation 11.6,

$$q_d = (110 - 62.4) \times \frac{8}{2} \times 23 = 4379 \text{ psf}$$

Note that in this application the soil density below the foundation is taken as the *effective density*, $\gamma - \gamma_w$, if the water table is relatively near the footing base. N_γ is determined from Figure 11.4 by determining the internal friction angle ϕ associated with the standard penetration resistance, and then the value of the factor to which it corresponds. The results in this case indicate that the footing is very near failure under the load shown.

11.2.2 Deep Foundations

Axially loaded piling and piers are generally supported by soil resistance at the tip and along the sides of the embedded portion. This distribution of load is indicated schematically in Figure 11.5. The total capacity of the piling

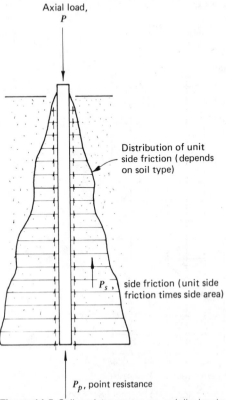

Axial load, P

Distribution of unit side friction (depends on soil type)

P_s, side friction (unit side friction times side area)

P_p, point resistance

Figure 11.5 Soil resistance on an axially loaded pile or pier.

Q_d is the sum of the capacities resulting from side friction and point resistance, as indicated by Equation 11.7:

$$Q_d = Q_s + Q_p \qquad (11.7)$$

In some situations, support may be derived either principally from side friction or principally from point resistance. In these situations, the structures are referred to as being either *friction* or *point-bearing* elements, respectively. Point-bearing piles or piers are usually economical where surface soils are soft and a firm supporting stratum exists a reasonable distance from the ground surface. Depths in excess of 100 ft are not uncommon. Friction piles are used when a satisfactory tip-bearing stratum cannot be economically reached. In the usual application, friction piles are installed to comparatively shallow depths.

In some applications, deep foundations must carry lateral load from a structure. Two methods of installation are used in this application. They are illustrated in Figure 11.6. Batter piles are inclined so that the resultant force on the pile head acts along the pile axis. The lateral force and the vertical force on the pile are in the same proportion as the lateral-to-vertical pile batter. In those situations where the lateral design loads are only moderate, analyses may be made for the lateral capacity and deflection of a vertical pile or pier. From the contractor's viewpoint, piles driven vertically are, of course, preferable to piles driven on a batter, because of the ease of construction involved. Piers are almost universally installed vertically.

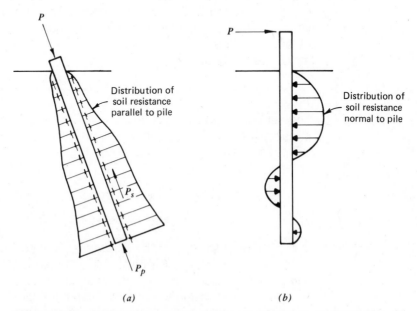

Figure 11.6 Soil resistance on laterally loaded piles and piers. *(a)* Batter piles. *(b)* Horizontal load on vertical pile or pier.

Deep foundations carry load as a result of their capability to develop point resistance and skin friction. How much resistance is available may be estimated by several methods. Examples of these methods are outlined in this section not, to present a state-of-the-art treatment of the subject, but to acquaint the reader with the fundamental considerations in deep foundation design.

The total force available for the development of skin friction and point resistance, as illustrated in Figure 11.5, can be estimated if certain soil properties are known and the dimensions of the piling or pier are given. Unit values of side frictional resistance are multiplied by the side area. Unit values of tip resistance are multiplied by the end area. The sum of the values so determined is the ultimate capacity. This is divided by a suitable factor of safety to arrive at an allowable load. This factor of safety is necessary to account for uncertainties in the analytical method and uncertainties in loads that may be placed on the piling. Calculation of the capacity of a friction pile driven in clay is illustrated by the following example:

Suppose that a 12-in. square prestressed concrete pile is to be driven 40 ft into a uniform deposit of clay having an unconfined compressive strength q_u of 2000 psf. Equation 11.7 may be expanded as

$$Q_d = Q_s + Q_p$$
$$= A_s f_s + A_p q_p$$

where f_s and q_p are unit values of side friction and point resistance, respectively. A_s and A_p are side and point areas. It has been shown that for soft clays, f_s is approximately equal to c, the undrained shear strength of the soil along the shaft. For circular foundations at great depth on clays, from Equation 11.3,

$$q_d = \gamma D_f + 9c$$

The net capacity of the tip q_p is $q_d - \gamma D_f$, since the increase in ultimate capacity resulting from depth is offset by the weight of the foundation element. Therefore, $q_p = 9c$ where c is the cohesion of the soil beneath the tip and

$$Q_d = 40 \times 4 \times 1 \times 1000 + 1 \times 1 \times 9 \times 1000$$
$$= 160,000 + 9000 = 169,000 \text{ lb}$$

With a factor of safety of 3 the allowable pile load would be about 56 kips or 28 tons. In this particular case, the concrete pile has much greater capacity as a structure than the soil can develop. It would likely be more economical to redesign the pile using timber if circumstances permitted.

In general, for deep foundations on clays, parameters for capacity analysis are selected as indicated by Table 11.2.

TABLE 11.2   SUMMARY OF PARAMETERS FOR BEARING-CAPACITY ANALYSIS OF PILES OR PIERS ON CLAYS

Point Capacity	Side Friction		
	Cohesion, c (tsf)	f_s/c	f_s
$q_p = 9c$	<0.25	1	c
	0.25–0.75	1–0.5	$1.0c$–$0.5c$
	>0.75	0.5	$0.5c$

For piles or piers on sand, the general methods outlined in the foregoing paragraphs apply, except that the values of side friction and tip bearing are selected differently. For side friction,

$$f_s = K_{p_0} \tan \phi' \tag{11.8}$$

where K is the coefficient of lateral earth pressure and p_0 is the vertical effective stress at the depth under consideration. Since p_0 increases with depth, calculations of capacity must be based on a summation of side friction forces over the length of the shaft. For point resistance, Equation 11.6 applies. For analysis of piles, the first term is usually neglected because it is a small part of the total load. The bearing-capacity factors are different for deep foundations than shallow foundations and may be selected from Table 11.3. Effective density should be used in all calculations, where applicable. The density associated with the first term is that below the foundation tip. In the second term, the quantity γD_f should be replaced by p_0, calculated at the depth of the tip. Research has shown that p_0 usually is less than that calculated from soil weight and, as a consequence, unless some adjustment is made, Equation 11.6 results in an overestimate of capacity.

To illustrate the application of the analysis, assume a 4-ft-diameter pier has been installed to a depth of 40 ft through clay having $c = 1000$ psf. Its tip is on water-bearing sand, for which $\phi' = 30°$ and $\gamma = 125$ pcf. The

TABLE 11.3   SUMMARY OF PARAMETERS FOR BEARING-CAPACITY ANALYSIS OF PILES AND PIERS ON SANDS

Point Capacity			Side Friction
Internal Friction Angle, ϕ' (degrees)	N_γ	N_q	$0.5 < K < 1.0$
35	100	100	For piers $K \leqslant 0.5$.
30	40	40	For displacement piles K.
25	18	20	may exceed 1.0 depending on installation method.

density of the clay is 110 pcf. From Equations 11.6 and 11.7,

$$Q_d = Q_s + Q_p$$
$$= A_s f_s + A_p q_p$$
$$= A_s f_s + A_p \left(\frac{\gamma B}{2} N_\gamma + p_0 N_q \right)$$

Now, from Table 11.2, $f_s = 750$ psf. Therefore,

$$Q_d = 4\pi \times 40 \times 750 + \frac{\pi}{4}$$
$$\times 4^2 \left[\frac{(125 - 62.4) \times 4}{2} \times 40 + 40 \times 110 \times 40 \right]$$
$$= 377{,}000 + 2{,}274{,}600 = 2{,}651{,}600 \text{ lb}$$
$$= 1326 \text{ tons}$$

The allowable load on this pier would be about 440 tons and should be confirmed by more refined analysis or load testing.

The most reliable method of determining capacity of a deep foundation is by test loading. Test loads are often required in a construction contract to overcome the uncertainties associated with the determination of allowable capacity. Load tests are justified, however, usually only on large jobs. The

Figure 11.7 Pile load test conducted by jacking against reaction piles.

load test method involves placing a load on the pile or pier to be tested either by stacking weights on a platform attached thereto, or by jacking using adjacent foundations for reactions. The latter method is illustrated in Figure 11.7. Load tests are generally of two types: proof tests or tests to failure. Proof tests are used simply to confirm a design assumption. For instance, the design may call for piles having a minimum allowable bearing capacity of 100 tons with a factor of safety of 2. Therefore, the required minimum ultimate capacity of the piling would be 200 tons. A proof test would involve measuring the deflection of the pile head at the 200-ton load. If this deflection were satisfactory, the pile would be acceptable. A test load to failure is used as a means of obtaining actual ultimate capacity or unit values of soil side friction and point resistance. These values obtained from actual load tests may be used to design piling of greater or lesser length or of different cross section than the actual test piling. A typical test load result, with the failure load indicated, is shown in Figure 11.8. Since incorporation of this information in the project may require some lead time, tests to failure generally will be done before the beginning of a project and may be a separate contract unto themselves.

There are various procedures for running pile load tests, depending on the particular information to be gained. A standard test method is specified in ASTM D1143.

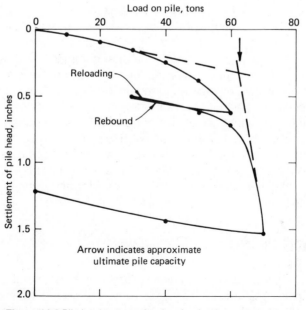

Figure 11.8 Pile load test results showing load versus settlement of pile head.

11.3 FOOTING AND RAFT CONSTRUCTION

Specifications usually require that groundwater be controlled in an excavation where shallow foundations are to be built, and that the footings be placed on undisturbed earth. This is done so that the as-built foundations bear on soils that have the same properties as those used in design. In some cases foundation soils will be improved by compaction or by replacement with better quality material.

Machine excavation usually is done with a mechanical shovel, backhoe, or tracked vehicle. Grading with such equipment loosens the upper few inches of supporting soil. The example in Section 11.1.1 shows that the major portion of the settlement beneath a footing is seated in the soil closest to the footing base. The loosened soil, which is more compressible than the natural soil, must, therefore, be removed. Hand excavation is often required to do so. Foundation construction in progress is shown in Figure 11.9.

Uncontrolled water in an excavation, either on the surface or seeping upward as a result of inadequate dewatering, will also loosen the surface soils. Water must, therefore, be controlled by sumping and dewatering. In the event that softening takes place, the unsuitable soil should be removed before construction of footings begins. The practice of dumping imported granular materials on the base of an excavation to provide trafficability for equipment is most often unacceptable, since the softened soil usually fills the voids in the coarser materials, resulting in a base that has nearly the same undesirable high compressibility.

It is a common practice to require that foundations and slabs be constructed on a thin lift of compacted sand or gravel. The usual purpose is to provide a highly pervious material communicating with a foundation drain system. In other cases this is done to provide material with limited compressibility immediately beneath footings. In both cases the final result desired is a clean, dense imported layer. Excavation in preparation for placing the layer must usually be done with the same care as that for foundations. Loosened material not removed will contaminate the coarser imported soil more easily than the undisturbed subgrade, particularly during compaction. If very coarse gravels are to be used, a thin layer of well-graded sand is sometimes placed first, to act as a filter and prevent intrusion of the natural soil.

11.4 FOUNDATION PILING

The principal use of driven piling in the context of this chapter is as load-bearing structural elements. They are used to transmit load either vertically or laterally to a satisfactory soil bearing stratum. On occasion, piles are also used to compact loose soils. In harbors and around ship berths, piles are

Figure 11.9 Footing foundation construction for a multistory building.

used for fendering systems and dolphins to protect more permanent structures.

The use of compaction piles has been discussed in Chapter 8. Such piles
are driven in loose cohesionless soil deposits to occupy void space and,
thereby, densify the material. They serve no structural purpose whatsoever.
Since compaction piles are designed to carry no load, and since they are
usually driven to a predetermined depth, the specifications are not particularly restrictive. Therefore, many of the problems associated with foundation
pile installation do not arise.

11.4.1 Pile Types

Depending on the basis of classification used, it would be possible to identify scores of different types of foundation piling. Considering the materials used, however, there are only four principal types. These are illustrated in Table 11.4.

The three principal materials involved are timber, concrete, and steel. Composite piles are made by combining two of these materials. No one pile type is universally accepted as being the best. Each has its own particular advantages and disadvantages that determine its suitability for a particular project. From a design viewpoint, steel piles and concrete piles generally are used to carry the heaviest load. Timber piles are limited in their carrying capability by the natural size of the tree from which the pile was cut. It has become an increasingly prevalent practice to permit design loads on timber piles up to 50 tons. The actual maximum will be based on the allowable stress for the wood and the pile size and load-carrying mode. The load-carrying capacity of steel and concrete piles is limited only by the size of section selected and the allowable stress used. This presumes, of course, that there is adequate soil support to develop this capacity.

It is a well-known fact that untreated timber piling installed full length below the permanent water table may be considered as permanent. Untreated piles above the water table are prone to decay as a result of bacteriological action. Creosote treating of timber piles alleviates this problem. Steel piles are susceptible to corrosion in the proper chemical environment. An abundant supply of oxygen and dissolved salts contributes to their deterioration. Studies have shown that the oxidation rate of steel piles driven full length in the ground is most rapid in that portion of the ground that is made up of recent fill. Concrete piles are subject to attack by acids that may be dissolved

TABLE 11.4   BASIC PILING TYPES

Material	Advantages	Limitations
Timber	Cutting, handling ease.	Damage during hard driving.
Precast concrete	Availability.	Length changes difficult. Handling care required. Harder driving than other high-capacity piles.
Cast-in place concrete (steel shells)	Adjustable length.	Two-step construction sequence.
Steel H-section or steel pipe	Adjustable length. Easy driving. Easy handling.	Economy. Availability.
Composite	Advantages of each material.	Disadvantages of each material. Three-step construction operation.

in the ground. Certain acid-resistant cements may be used to retard acid attack. On the whole then, when selecting a particular pile type for a given application, one must consider not only the load to be carried but also the environment in which the pile is to be installed.

From the contractor's viewpoint, the driveability of a piling and the ease with which its length may be adjusted are probably more important considerations than merely its capability to carry load. Piles that displace a large volume of soil as they are installed are more difficult to drive than *nondisplacement piles*. One could compare, for instance, a steel H-piling and a steel pipe piling driven with a closed end. The former will drive much more easily than the latter if the same hammer is used, even considering the possibility that the piles may be made of the same material and have the same material cross-sectional area. Concrete piles and timber piles are also considered *displacement piles*. Displacement piling then are particularly appropriate for use as friction piles. Driving displacement piling to great depth for point bearing may be difficult.

In many applications, the length of a driven piling cannot be determined prior to installation. In those cases where the bedrock surface has not been well defined by the exploratory drilling program, the length of point-bearing piles may vary considerably from estimates. In this situation, steel piles offer the advantage of relative ease in the adjustment of length. Extra sections may be added or extra lengths may be removed readily in the field. To alter the lengths of concrete piles by either splicing or cutting is a difficult field operation. The possibility that these adjustments in length may be required in a particular application should be considered by the contractor in preparing his bid. Traditionally, it has been the practice with timber piles to order extra length and cut off the portion remaining after driving was completed. This was permitted by the relative ease of making the cutoff and the comparative low cost of the timber material. Because material costs have risen, this practice is no longer a desirable alternative to accurate predetermination of driven lengths, particularly on large jobs.

Concrete piles may be cast in practically any size according to the particular application for a job. Steel piles are available in standard rolled sections such as those indicated in Figures 11.10 and 11.11. Timber pile sections are standard within ranges. These ranges are specified in ASTM D25 and are shown in Figure 11.12.

11.4.2 Pile Hammers

A pile-driving rig consists essentially of a set of leads, the hammer, and a crane as shown in Figure 11.13. The function of the leads is to guide the hammer and the pile during driving. In Figure 11.13 the leads are fixed (attached at the tip of the crane's boom), but the attachment is arranged so that the leads can be raised or lowered by sliding at the point of attachment. There is a horizontal brace or strut connecting the leads and crane.

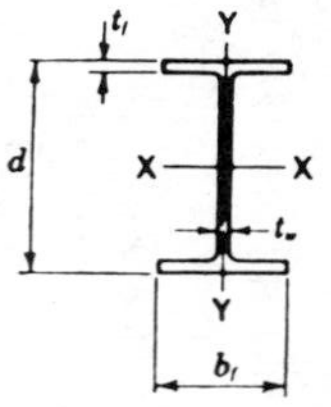

HP SHAPES
PROPERTIES FOR DESIGNING

			Flange		Web	Elastic Properties					
						Axis X-X			Axis Y-Y		
Designation	Area A	Depth d	Width b_f	Thickness t_f	Thickness t_w	I	S	r	I	S	r
	(in.²)	(in.)	(in.)	(in.)	(in.)	(in.⁴)	(in.³)	(in.)	(in.⁴)	(in.³)	(in.)
HP 14 × 117	34.4	14.21	14.885	0.805	0.805	1220	172	5.96	443	59.5	3.59
× 102	30.0	14.01	14.785	0.705	0.705	1050	150	5.92	380	51.4	3.56
× 89	26.1	13.83	14.695	0.615	0.615	904	131	5.88	326	44.3	3.53
× 73	21.4	13.61	14.585	0.505	0.505	729	107	5.84	261	35.8	3.49
HP 13 × 100	29.4	13.15	13.205	0.765	0.765	886	135	5.49	294	44.5	3.16
× 87	25.5	12.95	13.105	0.665	0.665	755	117	5.45	250	38.1	3.13
× 73	21.6	12.75	13.005	0.565	0.565	630	98.8	5.40	207	31.9	3.10
× 60	17.5	12.54	12.900	0.460	0.460	503	80.3	5.36	165	25.5	3.07
HP 12 × 84	24.6	12.28	12.295	0.685	0.685	650	106	5.14	213	34.6	2.94
× 74	21.8	12.12	12.217	0.607	0.607	566	93.4	5.10	185	30.2	2.91
× 63	18.4	11.94	12.125	0.515	0.515	472	79.1	5.06	153	25.3	2.88
× 53	15.5	11.78	12.045	0.435	0.435	393	66.8	5.03	127	21.1	2.86
HP 10 × 57	16.8	9.99	10.225	0.565	0.565	294	58.8	4.18	101	19.7	2.45
× 42	12.4	9.70	10.075	0.420	0.415	210	43.4	4.13	71.7	14.2	2.41
HP 8 × 36	10.6	8.02	8.155	0.445	0.445	119	29.8	3.36	40.3	9.88	1.95

Figure 11.10 Standard steel H-pile sections. (American Institute of Steel Construction.)

Size O.D.	Wall Thickness (in.)	Weight per linear ft (lb)	I Moment of Inertia (in.^{4})	r Radius of Gyration (in.)
8⅝″	.172	15.52	40.81	2.99
	.188	16.90	44.36	2.98
	.203	18.26	47.65	2.98
	.219	19.64	51.12	2.97
9″	.172	16.20	46.49	3.12
	.188	17.64	50.54	3.12
	.203	19.07	54.30	3.11
	.219	20.51	58.26	3.11
10″	.172	18.04	64.14	3.48
	.188	19.70	69.77	3.47
	.203	21.24	74.99	3.46
	.219	22.88	80.51	3.46
	.230	24.00	84.28	3.46
	.250	26.03	91.05	3.45
10¾″	.172	19.42	79.97	3.74
	.188	21.15	87.01	3.73
	.219	24.60	100.48	3.72
	.250	28.04	113.71	3.71
11″	.172	19.87	85.77	3.83
	.188	21.65	93.34	3.82
	.203	23.40	100.37	3.82
	.219	25.18	107.81	3.81
	.230	26.45	112.88	3.81
	.250	28.70	122.03	3.80
12″	.172	21.71	111.79	4.18
	.188	23.72	121.70	4.18
	.203	25.58	130.92	4.17
	.219	27.56	140.67	4.17
	.230	28.91	147.33	4.16
	.250	31.37	159.33	4.16
	.281	35.17	177.70	4.14
	.312	38.95	195.77	4.13
12¾″	.172	23.09	134.43	4.45
	.188	25.16	146.38	4.44
	.219	29.31	169.27	4.43
	.250	33.38	191.82	4.42
	.281	37.45	214.03	4.41
	.312	41.51	235.90	4.40
13″	.172	23.54	142.61	4.54
	.188	25.65	155.30	4.53
	.203	27.74	167.10	4.52
	.219	29.86	179.61	4.52
	.230	31.36	188.15	4.52
	.250	34.04	203.56	4.51
	.281	38.20	227.16	4.50
	.312	42.34	250.41	4.49

Figure 11.11 Dimensions and properties for standard steel pipe piling. Additional sizes are available. (Courtesy, L. B. Foster Company.) (continued next page)

Size O.D.	Wall Thickness (in.)	Weight per linear ft (lb)	I Moment of Inertia (in.⁴)	r Radius of Gyration (in.)
14″	.172	25.38	178.62	4.89
	.188	27.66	194.57	4.88
	.219	32.20	225.14	4.87
	.250	36.71	255.30	4.86
	.281	41.21	285.04	4.85
	.312	45.68	314.38	4.84
16″	.172	29.06	267.86	5.60
	.188	31.66	291.90	5.59
	.219	36.87	338.06	5.58
	.250	42.05	383.66	5.57
	.281	47.22	428.72	5.56
	.312	52.36	473.24	5.55
	.375	62.58	562.08	5.53
18″	.219	41.54	483.55	6.29
	.250	47.39	549.13	6.28
	.281	53.22	614.05	6.27
	.312	59.03	678.25	6.25
	.375	70.59	806.63	6.23

Figure 11.11 (continued)

On some setups the brace can be extended or shortened to incline the leads so that piles can be driven on a batter. For side batters the brace may be designed to move laterally, or it may have a moonbeam (upward curved beam supporting the lower end of the leads) at its end. Swinging leads are suspended from, rather than attached to, the boom. For offshore applications or in steep terrain, the pile may be supported by an external template and driven by a suspended hammer. No leads are used.

The essentials of a hammer-pile system are shown in Figure 11.14. A driving cap or helmet rests on the top of the pile to keep the hammer and pile together during driving. In driving concrete piles, a cushion, usually wood, is inserted between the pile head and the cap to prevent pile damage. In some circumstances, a cap block is used to cushion the hammer blow on the cap. The cap block frequently consists of a coiled length of wire rope, although other materials are used. The falling weight of the hammer strikes an anvil, a steel plate on top of the cap block. The anvil, cap block, and cushion may or may not be present according to the type of hammer and the piling being driven. There are five basic types of hammers. They are illustrated conceptually in Figure 11.15. Detailed specifications for a number of hammers are shown in Figure 11.16.

The drop hammer (see Figure 11.17) is the simplest hammer type and yet is not widely used because of the low available energy for driving and

Specified Butt Diameters, Inches	7	8	9	10	11	12	13	14	15	16	18
Required Minimum Circumference 3 ft from Butt	22	25	28	31	35	38	41	44	47	50	57
Length, Feet	Minimum Tip Circumferences (in.) and Corresponding Diameter in Parentheses										
20	16.0 (5.0)	16.0 (5.0)	16.0 (5.0)	18.0 (5.7)	22.0 (7.0)	25.0 (8.0)	28.0 (8.9)				
30	16.0 (5.0)	16.0 (5.0)	16.0 (5.0)	16.0 (5.0)	19.0 (6.0)	22.0 (7.0)	25.0 (8.0)	28.0 (8.9)			
40				16.0 (5.0)	17.0 (5.4)	20.0 (6.4)	23.0 (7.3)	26.0 (8.3)	29.0 (9.2)		
50						16.0 (5.0)	17.0 (5.5)	19.0 (6.0)	22.0 (7.0)	25.0 (8.0)	28.0 (8.9)
60						16.0 (5.0)	16.0 (5.0)	18.6 (5.9)	21.6 (6.9)	24.6 (7.8)	31.6 (10.0)
70						16.0 (5.0)	16.0 (5.0)	16.0 (5.0)	16.2 (5.1)	19.2 (6.1)	26.2 (8.3)
80							16.0 (5.0)	16.0 (5.0)	16.0 (5.0)	16.0 (5.0)	21.8 (6.9)
90							16.0 (5.0)	16.0 (5.0)	16.0 (5.0)	16.0 (5.0)	19.5 (6.2)
100							16.0 (5.0)	16.0 (5.0)	16.0 (5.0)	16.0 (5.0)	18.0 (5.8)
110										16.0 (5.0)	16.0 (5.0)
120											16.0 (5.0)

Note—Where the taper applied to the butt circumferences calculate to a circumference at the tip of less than 16 in., the individual values have been increased to 16 in. to assure a minimum of 5-in. tip for purposes of driving.

Figure 11.12 Standard timber pile sections. Select table according to specification requirement for butt or tip. (ASTM D25-70, reprinted, with permission, from the *Annual Book of ASTM Standards*. Copyright © 1970, American Society for Testing and Materials, 1916 Race Street, Philadelphia, PA. 19103.) (continued next page)

the high impact stresses it produces in piling. The concept of a drop hammer, however, is a useful background for the discussion of other hammer types. All hammers have a rated energy. The energy of a drop hammer is simply the potential energy of the raised weight above the pile head. The theoretical maximum amount of energy produced by a drop hammer is the product of the weight and the height of drop. Because of energy losses, the rated energy is multiplied by an efficiency factor to arrive at the energy available for driving.

A drop hammer is operated by lifting the weight with a cable and allowing it to fall freely, striking the head of the pile. This operation is cumbersome and time consuming and not suitable for high-production operations. A single-acting steam or air hammer operates on the same principle except that steam or air is used to raise the ram before it is dropped. At the top of a stroke, the chamber pressure raising the ram is exhausted and the ram falls. Single-acting hammers are widely used and particularly suited to driving heavy piles.

The rams for double-acting steam or air hammers are driven both on the up and down strokes. The ram is raised to the maximum height in much the

Specified Tip Diameter, Inches	5	6	7	8	9	10	11	12
Tip Circumference, Required Minimum	16	19	22	25	28	31	35	38
Length, Feet	Minimum Circumferences 3 ft from Butt (in.) with Diameter in Parentheses							
20	22.0 (7.0)	24.0 (7.6)	27.0 (8.6)	30.0 (9.5)	33.0 (10.5)	36.0 (11.5)	40.0 (12.7)	43.0 (13.7)
30	23.5 (7.5)	26.5 (8.4)	29.5 (9.4)	32.5 (10.3)	35.5 (11.3)	38.5 (12.2)	42.5 (13.5)	45.5 (14.5)
40	26.0 (8.3)	29.0 (9.2)	32.0 (10.2)	35.0 (11.1)	38.0 (12.1)	41.0 (13.0)	45.0 (14.3)	48.0 (15.3)
50	28.5 (9.0)	31.5 (10.0)	34.5 (11.0)	37.5 (11.9)	40.5 (12.9)	43.5 (13.8)	47.5 (15.1)	50.5 (16.0)
60	31.0 (9.8)	34.0 (10.8)	37.0 (11.8)	40.0 (12.7)	43.0 (13.7)	46.0 (14.6)	50.0 (15.9)	53.0 (16.8)
70	33.5 (10.6)	36.5 (11.6)	39.5 (12.6)	42.5 (13.5)	45.5 (14.4)	48.5 (15.4)	52.5 (16.7)	55.5 (17.7)
80	36.0 (11.4)	39.0 (12.4)	42.0 (13.4)	45.0 (14.3)	48.0 (15.3)	51.0 (16.2)	55.0 (17.5)	58.0 (18.4)
90	38.6 (12.2)	41.6 (13.2)	44.6 (14.2)	47.6 (15.1)	50.6 (16.0)	53.6 (17.0)	57.6 (18.3)	60.5 (19.2)
100	41.0 (13.0)	44.0 (14.0)	47.0 (15.0)	50.0 (15.9)	53.0 (16.8)	56.0 (17.8)	60.0 (19.0)	
110	43.6 (13.8)	46.6 (14.8)	49.6 (15.7)	52.6 (16.7)	55.6 (17.7)	61.0 (19.4)		
120	46.0 (14.6)	49.0 (15.6)	52.0 (16.6)	55.0 (17.5)	58.0 (18.4)			

Figure 11.12 Continued

same manner as a single-acting hammer. On the down stroke, steam or air is injected, which accelerates the downward movement of the ram. Accordingly, the maximum energy available is the sum of the potential energy resulting from the weight of the ram plus the additional energy imparted by the injection at the top of the downstroke. Double-acting hammers are widely used for driving light piling in relatively easy driving conditions. They operate rapidly. It should be pointed out that a double-acting hammer may develop the same rated energy as a comparable single-acting hammer in one of two ways. If the weight of the ram in the two hammers were the same, they must have the same velocity at impact to have the same kinetic energy. This requires that the height of fall of the double-acting hammer be lower than for the comparable single-acting hammer of the same weight and that the additional energy to accelerate the ram be made up by injection of gas at the top of the stroke. However, if the ram weight for the double-acting hammer is less than that of the corresponding single-acting hammer, the ram of the double-acting hammer must travel at a higher velocity at impact to achieve the same energy. This additional velocity must be achieved through the use of either an increased height of drop, increased speed due to gas injection, or a combination of both. These matters are particularly important when considering potential damaging effects of hammers on piling. It can be shown, for instance, that the stresses produced in a pile by driving are proportional to the velocity of the hammer at impact. Single-

Figure 11.13 Driving steel pipe piling for a large structure foundation with a single-acting steam hammer.

acting hammers and double-acting hammers of the same rated energy, therefore, are not necessarily equally suited to the driving of a particular pile.

Diesel hammers (see Figure 11.18) have achieved increasing popularity

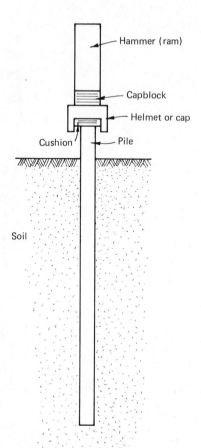

Figure 11.14 Schematic of hammer-pile system.

in recent years. They offer the advantage of being entirely self-contained, requiring no external boiler or compressor system for operation. Combustion of diesel is produced as the result of the compression of a fuel-air mixture in the space below the falling ram. A diesel hammer must, therefore, be started manually on the first stroke. For each succeeding stroke, the hammer runs itself since the explosion raises the ram and permits it to fall, compressing the gas for the next ignition. The rated energy of a diesel hammer is a combination of the potential energy produced by the height of drop of the ram and the additional energy imparted by the resulting fuel combustion. Double-acting diesel hammers are available. Diesel hammers operate at rated energy generally when hard driving resistance is encountered. In softer driving, the lower resistance does not permit the ram to recoil for its full height of drop, and the operation is, therefore, slower and results in less energy. In very soft ground conditions, diesel hammers will sometimes fail to operate properly since the pile does not offer enough resistance to penetration to permit the needed compression for ignition to occur.

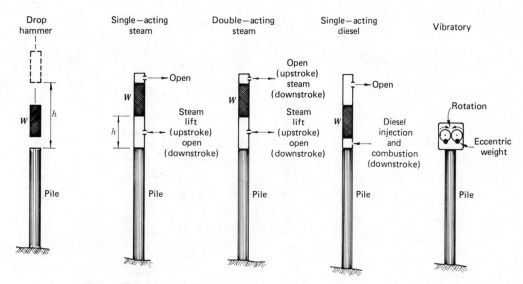

Figure 11.15 Conceptual representation of pile hammers.

Vibratory hammers (see Figure 11.19) are fundamentally different from the other types previously discussed. These hammers transmit a sinusoidally varying force to the top of the pile as a result of the centrifugal forces produced by a pair of eccentric weights operated in synchronization by electric motors in the hammer. Vibratory hammers are particularly suited for driving nondisplacement piling and sheet piling in sandy soils.

A number of factors bear on the selection of a particular hammer type for a particular job. Not the least of these factors is ownership or availability of the equipment. The pile size and type must be considered, as well as the driving energy required. In addition, the speed of the driving operation is an important factor in preparing the bid. For operation in adverse weather conditions or underwater, special hammer types are required. As a general rule, it is advantageous to drive piling with the hammer that is large enough to produce the maximum penetration rate and, at the same time, not damage the piles. Hammer selection has been, and continues to be, largely a matter of the experience of a particular contractor within the limits imposed on him by the contract specifications. Analytical approaches to hammer selection are in the continuing stages of development. They are based on techniques for computing pile capacity or driving stresses. These concepts are introduced in later sections of this chapter.

11.4.3 Jetting, Spudding, and Predrilling

In some circumstances, piling must be driven through dense or hard materials to bearing at a greater depth. Several methods for doing so have been developed. The principal functions of these driving aids are to speed

AIR OR STEAM HAMMERS

RATED ENERGY FT./LBS.	SIZE	MAKE	TYPE	STYLE	BLOWS PER MIN.	WT. OF STRIKING PARTS	TOTAL WEIGHT LBS.	HAMMER LENGTH	JAW DIMENSIONS	BOILER HP REQ'D ASME	STEAM CONSUMP. LBS./HR.	CFM REQ'D	PSI REQ'D	INLET SIZE	VEW RATING
632,885	MRBS 7000	MKT/MENCK	SGL. ACT.	OPEN	35	154,000	303,600	30'2"	CAGE	1,192	34,257	— —	142	(4)4"	312,193
600,000	4150-OT	VULCAN	SGL. ACT.	OPEN	50	150,000	295,000	31'7"	24"x112"	2,000	66,891	16,750	150	(2)6"	300,000
600,000	4150-CT	VULCAN	SGL. ACT.	OPEN	50	150,000	307,000	31'7"	24"x112"	2,000	66,891	16,750	150	(2)6"	300,000
450,000	3150-OT	VULCAN	SGL. ACT.	OPEN	60	150,000	289,000	28'7"	24"x112"	2,000	67,443	16,508	150	(2)6"	259,808
450,000	3150-CT	VULCAN	SGL. ACT.	OPEN	60	150,000	301,000	28'7"	24"x112"	2,000	67,443	16,508	150	(2)6"	259,808
361,650	MRBS 4000	MKT/MENCK	SGL. ACT.	OPEN	35	88,000	160,600	26'4"	CAGE	668	19,300	— —	142	(2)4"	178,396
300,000	560	VULCAN	SGL. ACT.	OPEN	45	62,500	134,060	23'0"	18¾"x88"	875	30,150	5,633	150	(2)5"	136,930
226,030	MRBS 2500	MKT/MENCK	SGL. ACT.	OPEN	35	55,000	98,600	23'4"	CAGE	426	12,235	— —	142	(2)3"	111,497
200,000	540	VULCAN	SGL. ACT.	OPEN	48	40,900	102,980	21'11"	14"x80"	635	21,900	4,200	130	(2)5"	89,000
180,000	OS-60	MKT		CLOSED	55	60,000	141,150	25'9"	14¾"x72"(M)	750	— —		135	(2)3"	103,923
180,000	360	VULCAN	SGL. ACT.	OPEN	62	60,000	124,830	19'0"	18¾"x88"	750	25,556	4,626	130	(2)4"	103,923
180,000	060	VULCAN	SGL. ACT.	OPEN	62	60,000	128,840	19'0"	18¾"x88"(M)	750	25,556	4,626(A) 8,995(I)	130	(2)4"	103,923
135,620	MRBS 1500	MKT/MENCK	SGL. ACT.	OPEN	35	33,000	58,080	18'6"	CAGE	260	7,540	— —	142	4"	66,898
120,000	OS-40	MKT	SGL. ACT.	CLOSED	55	40,000	111,000	21'3"	14¾"x72"(M)	530	— —		135	(2)3"	69,282
120,000	340	VULCAN	SGL. ACT.	OPEN	60	40,000	98,180	17'11"	14"x80"	600	18,446	3,400	120	(2)3"	69,282
120,000	040	VULCAN	SGL. ACT.	OPEN	60	40,000	88,000	19'3"	14¼"x50"	280	10,250	4,500	— —	5"	69,282
113,488	400-C	VULCAN	DIFFER.	OPEN	100	40,000	83,000	16'8"	14¼"x50"	700	24,150	4,659	150	5"	63,378
90,000	030	VULCAN	SGL. ACT.	OPEN	55	30,000	54,000	15'0"	11¼"x37"	300	9,000	3,000	150	3"	51,964
90,000	300	CONMACO	SGL. ACT.	OPEN	55	30,000	55,390	16'5"	11¼"x56"	247	8,550	1,883(A) 3,724(I)	150	3"	51,961
81,250	8/0	RAYMOND	SGL. ACT.	OPEN	35	25,000	34,000	19'4"	10¼"x25"	246	8,500	— —	135	3"	45,069
75,000	30X	RAYMOND	SGL. ACT.	OPEN	52	30,000	52,000	19'1"	— —	246	8,500	— —	135	3"	47,434
67,825	MRBS 750	MKT/MENCK	SGL. ACT.	OPEN	40	16,500	25,250	17'8"	CAGE	128	3,670	1,590	100-114	3"	33,453
60,000	S-20	MKT	SGL. ACT.	CLOSED	60	20,000	38,650	15'5"	*x36"	190	— —	1,720	150	3"	34,640
60,000	020	VULCAN	SGL. ACT.	OPEN	60	20,000	39,000	15'0"	11¼"x37"	217	7,500	1,634	120	3"	34,641
60,000	200	CONMACO	SGL. ACT.	OPEN	60	20,000	44,560	15'0"	11¼"x56"	217	7,486	1,634(A) 3,093(I)	120	3"	34,641
56,875	5/0	RAYMOND	SGL. ACT.	OPEN	44	17,500	26,450	16'9"	10¼"x25"	100	4,250	— —	150	3"	31,548
50,200	200-C	VULCAN	DIFFER.	OPEN	98	20,000	39,050	13'2"	11¼"x37"	260	8,970	1,746	142	4"	31,685
48,750	016	VULCAN	SGL. ACT.	OPEN	60	16,250	30,000	14'6"	11¼"x32"	210	6,950	1,282	120	3"	28,148
48,750	4/0	RAYMOND	SGL. ACT.	OPEN	46	15,000	23,800	16'1"	— —	85	— —	— —	120	2½"	27,042
48,750	150-C	RAYMOND	DIFFER.	OPEN	95-105	15,000	32,500	15'9"	— —	— —	— —	— —	120	3"	27,042
48,750	180	CONMACO	SGL. ACT.	OPEN	60	16,250	33,200	13'10"	11¼"x42"	198	6,950	1,930	120	3"	28,148
45,200	MRBS 500	MKT/MENCK	SGL. ACT.	OPEN	40	11,000	15,200	16'9"	CAGE	85	2,450	1,060	100-114	2½"	22,298
42,000	014	VULCAN	SGL. ACT.	OPEN	60	14,000	27,500	14'6"	11¼"x32"	200	6,920	1,200	110	3"	24,248
42,000	140	CONMACO	SGL. ACT.	OPEN	60	14,000	30,750	13'10"	11¼"x42"	179	6,185	1,875	110	3"	22,248
41,280	160D	CONMACO	DIFFER.	OPEN	103	16,000	35,400	13'7½"	11¼"x42"	237	8,175	1,865	160	3"	25,700
40,625	125	CONMACO	SGL. ACT.	OPEN	50	12,500	21,430	14'10"	9¼"x32"	120	4,120	1,635	125	2½"	22,534
40,600	3/0	RAYMOND	SGL. ACT.	OPEN	48	12,500	21,225	15'7"	10¼"x25"	— —	3,000	— —	120	2½"	22,528
37,500	S-14	MKT		CLOSED	60	14,000	31,700	13'7"	*x36"	155	— —	1,260	100	3"	23,000
37,375	115	CONMACO	SGL. ACT.	OPEN	50	11,500	20,780 / 20,250	14'2" / 15'0"	9¼"x32"(C) 9¼"x26"(K)	100 / 115	3,425 / 3,980	1,365 / 1,585	120	2½"	20,732
36,000	140-C	VULCAN	DIFFER.	OPEN	103	14,000	27,984	13'2"	11¼"x32"	211	7,279	1,425	140	3"	22,449
36,000	140D	CONMACO	DIFFER.	OPEN	103	14,000	31,200	12'3"	11¼"x42"	211	7,279	1,650	140	3"	22,449
32,500	2/0	RAYMOND	SGL. ACT.	OPEN	50	10,000	18,550	15'0"	10¼"x25"	— —	2,400	— —	110	2"	18,028
32,500	010	VULCAN	SGL. ACT.	OPEN	50	10,000	18,750	15'0"	9¼"x26"	157	5,440	1,002	105	2½"	18,028
32,500	100	CONMACO	SGL. ACT.	OPEN	50	10,000	19,280 / 18,700	14'2" / 15'0"	9¼"x32"(C) 9¼"x26"(K)	85 / 100	2,945 / 3,425	1,170 / 1,360	100	2½"	18,028
32,500	S-10	MKT	SGL. ACT.	CLOSED	55	10,000	22,380	14'1"	*x30"	130	5,440	1,000	80	2½"	18,028
26,000	08	VULCAN	SGL. ACT.	OPEN	50	8,000	16,750	15'0"	9¼"x26"	127	4,380	880	83	2½"	14,422
26,000	S-8	MKT	SGL. ACT.	CLOSED	55	8,000	18,300	14'4"	*x26"	120	4,380	850	80	2½"	14,422
26,000	80	CONMACO	SGL. ACT.	OPEN	50	8,000	17,280 / 16,700	14'2" / 15'0"	9¼"x32"(C) 9¼"x26"(K)	75 / 87	2,580 / 3,000	1,010 / 1,175	85	2½"	14,422
24,450	80-C	VULCAN	DIFFER.	OPEN	111	8,000	17,885	12'2"	9¼"x26"	180	6,210	1,245	120	2½"	13,985
24,450	80-CH	RAYMOND	DIFFER.	OPEN	110-120	8,000	17,782	11'10"	— —	N/A	N/A	— —	— —	— —	13,985
24,450	80-C	RAYMOND	DIFFER.	OPEN	95-105	8,000	17,885	12'2"	— —	— —	— —	— —	135	2½"	13,985
24,375	0	RAYMOND	SGL. ACT.	OPEN	52	7,500	16,000	15'0"	10¼"x25"	— —	— —	750	110	2"	13,485
24,375	0	VULCAN	SGL. ACT.	OPEN	50	7,500	16,250	15'0"	9¼"x26"	155	4,380	841	80	2½"	13,485
24,000	C-826	MKT	COMPOUND	CLOSED	85-95	8,000	17,750	12'2"	*x26"	120	— —	875	125	2½"	13,856
19,500	65-C	RAYMOND	DIFFER.	OPEN	110	6,500	14,675	11'8"	9¼"x19"	— —	3,100	— —	120	2"	11,201
19,500	1-S	RAYMOND	SGL. ACT.	OPEN	58	6,500	12,500	12'9"	7½"x28¼"	— —	1,500	— —	104	1½"	11,258
19,500	06	VULCAN	SGL. ACT.	OPEN	60	6,500	11,200	13'0"	8¼"x20"	94	3,230	625	100	2"	11,258
19,500	65	CONMACO	SGL. ACT.	OPEN	60	6,500	12,100 / 11,200	13'0"	9¼"x26"(C) 8¼"x20"(K)	67	2,300	900	100	2"	11,258
19,500	65-CH	RAYMOND	DIFFER.	OPEN	130	6,500	14,615	12'1"	— —	N/A	N/A	— —	— —	2"	11,258
19,200	65-C	VULCAN	DIFFER.	OPEN	117	6,500	14,886	12'2"	9¼"x20"	152	5,244	991	150	2"	11,201
19,150	11B3	MKT	DBL. ACT.	CLOSED	95	5,000	14,000	11'2"	*x26"	126	— —	900	100	2½"	9,785
16,250	S-5	MKT	DBL. ACT.	CLOSED	60	5,000	12,460	13'3"	*x24"	85	— —	600	80	2"	9,000
16,000	C-5	MKT	DBL. ACT.	CLOSED	100-110	5,000	11,880	8'9"	*x26"	80	— —	585	100	2½"	8,944
15,100	50-C	VULCAN	DIFFER.	OPEN	120	5,000	11,782	11'0"	8¼"x20"	125	4,312	880	120	2"	8,689
15,000	1	VULCAN	SGL. ACT.	OPEN	60	5,000	9,700	13'0"	8¼"x20"	81	2,794	565	80	2"	8,660
15,000	1	RAYMOND	SGL. ACT.	OPEN	60	5,000	11,400	12'9"	7½"x28¼"	— —	1,400	500	80	1½"	8,660
15,000	50	CONMACO	SGL. ACT.	OPEN	60	5,000	10,600 / 9,700	13'0"	9¼"x26"(C) 8¼"x20"(K)	56	1,925	740	80	2"	8,660
13,100	10B3	MKT	DBL. ACT.	CLOSED	105	3,000	10,850	9'2"	**x24"	104	— —	750	100	2½"	6,269
8,750	9B3	MKT	DBL. ACT.	CLOSED	145	1,600	7,000	8'4"	8½"x20"	85	— —	600	100	2"	3,742
7,260	30-C	VULCAN	DIFFER.	OPEN	133	3,000	7,036	9'8"	7¼"x19"	70	2,412	488	120	1½"	4,666
7,260	2	VULCAN	SGL. ACT.	OPEN	70	3,000	6,700	11'6"	7¼"x19"	49	1,690	336	80	1½"	4,666
4,150	7	MKT	DBL. ACT.	CLOSED	225	800	5,000	6'1"	*x21"	65	— —	450	100	1½"	1,697
4,000	DGH-900	VULCAN	DIFFER.	CLOSED	238	900	5,000	6'9"	VARIES	40	— —	580	78	1½"	1,897
2,500	6	MKT	DBL. ACT.	CLOSED	275	400	2,900	5'3-1/8"	*x15"	45	— —	400	100	1¼"	1,000
1,000	5	MKT	DBL. ACT.	CLOSED	300	200	1,500	4'7"	6"x11"	35	— —	250	100	1¼"	447
386	DGH-100A	VULCAN	DIFFER.	CLOSED	303	100	786	4'2"	4¼"x8¾"	5	— —	74	60	2"	196
356	3	MKT	DBL. ACT.	CLOSED	400	68	675	4'10"	NONE	25	— —	110	100	1"	155
	2	MKT	DBL. ACT.	CLOSED	500	48	343	2'9"	NONE	15	— —	70	125	¾"	— —
	1	MKT	DBL. ACT.	CLOSED	500	21	145	3'7"	NONE	15	— —	70	125	¾"	— —

Figure 11.16 Specifications for pile hammers. (Conmaco, Inc., Kansas City, Kansas.)

DIESEL HAMMERS

RATED ENERGY FT./LBS.	MODEL	MANUFACTURER	SINGLE/DOUBLE ACTING	OPEN/CLOSED TOP	BLOWS MIN.	PISTON WEIGHT	MAX. STROKE	TOTAL WEIGHT	TOTAL LENGTH	JAW SIZE	FUEL USED GPH	VEW
280,000	K150	KOBE	SGL. ACT.	OPEN	45-60	33,100	8'6"	80,500	29'8"	CAGE	16-20	92,870
137,000	MB70	MITSUBISHI	SGL. ACT.	OPEN	38-60	15,840	8'8"	46,000	19'6"	— —	7-10	46,584
117,175	D55	DELMAG	SGL. ACT.	OPEN	36-47	11,860	9'10"	26,300	17'9"	—x32"	5.54	37,279
105,600	K60	KOBE	SGL. ACT.	OPEN	42-60	13,200	8'0"	37,500	24'3"	9"x42"	6.5-8.0	37,335
91,100	K45	KOBE	SGL. ACT.	OPEN	39-60	9,900	9'2"	25,600	18'6"	9"x36"	4.5-5.5	30,031
87,000	D44	DELMAG	SGL. ACT.	OPEN	37-56	9,500	9'2"	22,300	15'8"	—x32"	4.5	28,749
84,000	M43	MITSUBISHI	SGL. ACT.	OPEN	40-60	9,460	8'10"	22,660	16'3"	11¼"x37"	4.0-5.8	28,189
79,500	J44	MKT/IHI	SGL. ACT.	OPEN	42-70	9,720	8'2"	21,500	14'10"	—x37"	6.86	27,798
79,000	K42	KOBE	SGL. ACT.	OPEN	40-60	9,260	8'6"	24,000	17'8"	—x36"	4.5-5.5	27,047
73,780	D36	DELMAG	SGL. ACT.	OPEN	37-53	7,940	9'3"	17,780	14'11"	—x32"	3.7	24,204
70,800	K35	KOBE	SGL. ACT.	OPEN	39-60	7,700	9'2"	18,700	17'8"	9"x30"	3.0-4.0	23,349
64,000	M33	MITSUBISHI	SGL. ACT.	OPEN	40-60	7,260	8'0"	16,940	13'2"	9¼"x32"	3.4-5.3	21,556
63,500	J35	MKT/IHI	SGL. ACT.	OPEN	42-70	7,730	8'3"	16,900	14'6"	11"x32"	4.76	22,155
60,100	K32	KOBE	SGL. ACT.	OPEN	40-60	7,050	8'6"	17,750	17'8"	9"x30"	2.75-3.5	20,584
56,000	DE70	MKT	SGL. ACT.	OPEN	48	7,000	12'4"	12,250	15'10"	—x26"	5.0	19,799
54,250	D30	DELMAG	SGL. ACT.	OPEN	39-60	6,600	8'3"	12,160	14'2"	—x26"	2.9	18,922
50,700	K25	KOBE	SGL. ACT.	OPEN	39-60	5,510	9'3"	13,100	17'6"	9"x26"	2.5-3.0	16,714
45,000	M23	MITSUBISHI	SGL. ACT.	OPEN	42-60	5,060	8'10"	11,200	14'1"	9¼"x26"	2.4-3.7	15,090
43,400	N60	VULCAN	SGL. ACT.	OPEN	50-60	5,280	8'2"	12,760	15'1"	11¼"x26"	1.85	15,138
41,300	K22	KOBE	SGL. ACT.	OPEN	40-60	4,850	9'2"	12,350	17'6"	9"x26"	2.0-2.75	14,153
40,000	DA55B	MKT	SGL. ACT.	OPEN	48	5,000	10'9"	18,300	17'4"	—x26"	4	14,142
39,700	D22	DELMAG	SGL. ACT.	OPEN	42-60	4,850	8'2"	11,150	14'2"	—x26"	3.44	13,876
39,100	J22	MKT/IHI	SGL. ACT.	OPEN	42-70	4,850	10'0"	10,800	14'0"	9¼"x26	3.2	13,771
38,000	DA55B	MKT	DBL. ACT.	CLOSED	80	5,000	7'6"	18,300	17'4"	—x26"	4.0	13,784
32,549	N46	VULCAN	SGL. ACT.	OPEN	50-60	3,960	8'2"	9,845	15'1"	9¼"x26	1.59	11,353
32,000	DE40	MKT	SGL. ACT.	OPEN	48	4,000	10'9"	9,825	15'0"	—x26"	3.0	11,314
30,000	520	LINK BELT	DBL. ACT.	CLOSED	80-84	5,070	62.19"	12,545	13'6"	—x26"	1.35	12,333
27,100	D15	DELMAG	SGL. ACT.	OPEN	42-57	3,300	8'3"	6,650	13'11"	—x20"	1.75	9,457
26,000	M14S	MITSUBISHI	SGL. ACT.	OPEN	42-60	2,970	8'9"	7,260	13'7"	9¼"x26	1.3-2.1	8,787
24,600	N33	VULCAN	SGL. ACT.	OPEN	50-60	3,000	8'2"	7,645	15'8"	9¼"x26	1.32	8,591
24,400	K13	KOBE	SGL. ACT.	OPEN	40-60	2,860	8'6"	7,300	16'8"	9"x26"	.75-2.0	8,354
22,500	D12	DELMAG	SGL. ACT.	OPEN	42-60	2,750	8'2"	6,050	13'11"	—x20"	2.11	7,866
22,400	DE30	MKT	SGL. ACT.	OPEN	48	2,800	10'9"	8,125	15'0"	—x20"	2.0	7,920
22,400	DA35B	MKT	SGL. ACT.	OPEN	48	2,800	10'9"	10,000	17'0"	—x20"	2.0	7,920
21,000	DA35B	MKT	DBL. ACT.	CLOSED	82	2,800	—	10,000	17'0"	—x20"	2.7	7,668
18,000	440	LINK BELT	DBL. ACT.	CLOSED	86-90	4,000	54.6"	10,300	14'6"	—x20"	1.6	8,532
18,000	312	LINK BELT	DBL. ACT.	CLOSED	100-105	3,857	46.41"	10,375	10'9"	—x26"	1.1	8,332
16,000	DE-20	MKT	SGL. ACT.	OPEN	48	2,000	9'5"	5,375	13'3"	—x20"	1.6	5,657
9,100	D-5	DELMAG	SGL. ACT.	OPEN	42-60	1,100	8'3"	2,730	12'2"	—x19"	1.32	3,164
8,800	DE-10	MKT	SGL. ACT.	OPEN	48	1,100	8'0"	3,100	12'2"	10" BEAM	.9	3,111
8,100	180	LINK BELT	DBL. ACT.	CLOSED	90-95	1,725	55.68"	4,550	11'3"	8"x18"	.65	3,738
3,630	D-4	DELMAG	SGL. ACT.	OPEN	50-60	836	4'4"	1,360	7'9"	SPUD CLIP	.21	1,742
1,815	D-2	DELMAG	SGL. ACT.	OPEN	60-70	484	4'1"	792	6'9"	SPUD CLIP	.075	937

Figure 11.16 (continued)

Figure 11.17 A lightweight drop hammer used for installation of timber piles. Helmet is shown suspended on hammer above pile head.

the driving operation and to prevent damage to piling that results from heavy driving. Jetting is applicable to those situations where piling must be driven through cohesionless soil materials to greater depths. Jetting is accomplished by pumping water through pipes attached to the side of the pile as it is driven. The flow of water creates a *quick* condition and thereby reduces skin friction along the sides of the pile. The result is that the pile drives more easily. Where piles have been designed to carry a large portion of their load in skin friction, jetting may not be permitted, since it loosens the soil and thereby could reduce pile capacity.

Spudding and predrilling are both operations that produce a hole into which a driven piling may be inserted. Spudding is accomplished by driving a heavy steel mandrel that is subsequently removed. Spudding is particularly

Figure 11.18 Diesel pile hammer.

applicable to installing piling through materials containing debris and large rocks. Predrilling accomplishes the same results except that an auger, rather than a mandrel, is used. In both cases, the stratum to be spudded or predrilled must be cohesive enough so that the resulting hole remains open until pile driving begins.

Figure 11.19 A vibratory pile hammer used to drive steel sheetpiling.

11.4.4 Dynamic Formulas

The method of pile capacity calculation most often encountered by the contractor involves the use of dynamic pile-driving formulas. Such formulas are employed in field control of driving operations. The concept is illustrated by reference to Sander's formula. In Sander's formula, it is assumed that the work done in driving a piling may be equated to the energy available for driving:

$$Wh = Q_d s \qquad (11.9)$$

W is the ram weight, h is the height of drop, Q_d is the capacity, and s is the penetration of the pile resulting from one hammer blow. In this formula, the driving energy for a drop hammer or a single-acting steam hammer is equated to the work done in moving the pile. That is, the hammer's potential energy is equated to the product of a force, which is the resistance of the piling to penetration, and the distance through which the pile moves. Solving this relationship for capacity produces Equation 11.10, which is the simplest of pile-driving formulas:

$$Q_d = \frac{Wh}{s} \qquad (11.10)$$

Its origin may be traced back to about 1850. It suggests the attractive possibility that the capacity of the piling can be determined by knowing the energy of the hammer and measuring the penetration of the piling resulting from a hammer blow. Most of the several hundred dynamic pile-driving formulas that are now available are based on modifications of this basic approach that have been developed in various attempts to overcome its inadequacies.

The most widely used of these formulas is the *Engineering News* formula. Equation 11.11 is Sander's formula modified to indicate that had not certain energy losses occurred, the piling would have penetrated an additional amount, c.

$$Q_d = \frac{Wh}{s + c} \qquad (11.11)$$

If we measure s and c in inches, and h in feet, the equation is dimensionally correct in the form

$$Q_d = \frac{12Wh}{s + c} \qquad (11.12)$$

Equation 11.12 theoretically defines the ultimate capacity of the piling. The *Engineering News* formula was suggested for use with the factor of safety of 6. Accordingly, the allowable capacity of a piling would then be computed from

$$Q_a = \frac{2Wh}{s + c} \qquad (11.13)$$

The suggested value of c is 1 in. for drop hammers and 0.1 for steam hammers.

The preceding development was offered not to indicate the suitability of the *Engineering News* formula, but rather to illustrate the principles on which

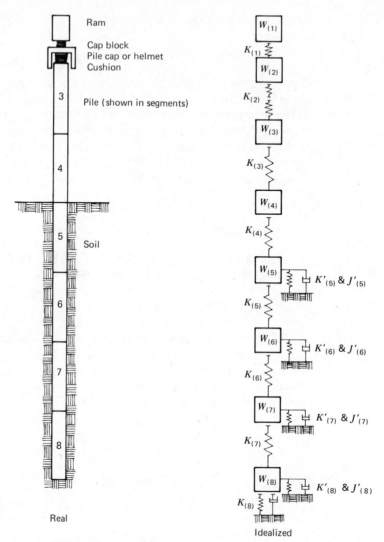

Figure 11.20 Pile-hammer representation for wave equation analysis.

it and many other driving formulas are based. It is generally known that this equation is an inadequate method of determining actual pile-bearing capacities. Nonetheless, it is widely used to control field-driving operations in the United States.

At the other end of the spectrum of dynamic formulas is the method of wave equation analysis. In this method, the pile and hammer are treated not as rigid bodies, but as linearly elastic rods represented by a series of springs and weights as shown in Figure 11.20. With proper substitution of appropriate properties of the hammer, pile, and soil, it has been repeatedly shown that the wave analysis method is the most reliable method of dynamic analysis available today. It offers the capability not only of good pile capacity predictions, but also a method of calculating stresses produced in piles by driving. Examples of such analyses are given in Figures 11.21 and 11.22. It therefore can be used as a reliable aid in the proper selection of a pile-driving hammer. The principal disadvantage associated with the wave analysis method lies with the need to use a digital computer for the associated calculations.

Penetration resistance versus depth plots for piles driven in sands and clays are shown in Figure 11.23. The units of penetration resistance shown, blows per foot, may be inverted to obtain s, in feet per blow, or inches per blow. We know from experience and the analytical methods presented earlier that the capacity of piles driven in homogeneous soils increases approximately linearly with depth. Yet according to Equation 11.13, the capacity of the pile indicated in Figure 11.23 as being driven in clay would remain

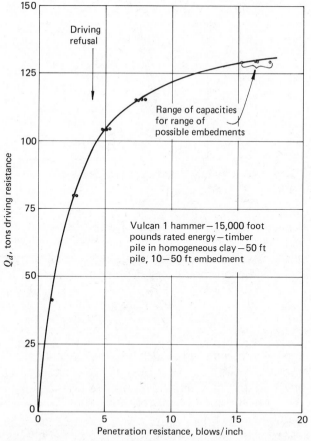

Figure 11.21 Dynamic capacity or driving resistance related to penetration resistance by wave equation analysis.

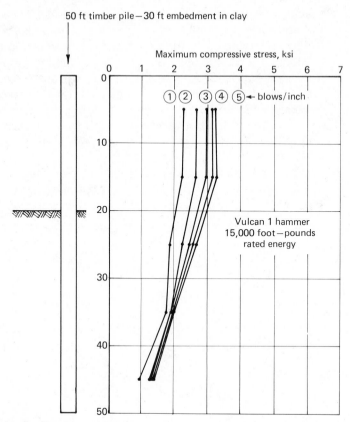

Figure 11.22 Compressive stresses in a timber pile during driving from wave equation analysis.

constant regardless of depth. These observations preclude the possibility that the usual pile-driving formulas may be reliably used to determine capacity of piles driven in homogeneous clay. They are, therefore, unsuited for construction control for this purpose. They work reasonably well in sands.

11.4.5 Contract Provisions

Various problems arise in pile construction because of the uncertainties involved in pile analysis in design. For instance, methods for estimating the driven length of friction piles are not well developed nor are methods of selecting the appropriate hammer. Furthermore, our ability to predict the load-carrying capacity of a piling is not as good as it could be. There are also problems associated with the external effects of the pile-driving operation. All of these uncertainties contribute to further uncertainties in bidding and in the payment for work actually accomplished. Accordingly, pile-driving contracts often produce disputes.

A driving contract includes the normal provisions for materials, workmanship, and payment. The particular pile material to be used is usually selected after careful consideration by the design engineer; however, the contractor in some instances is permitted the latitude of offering an alternate. For example, a contractor may be offered the alternative of driving prestressed concrete piling to tip bearing instead of steel H-piling in the case where the concrete piles may be easier to obtain. The contractor must recognize, however, that the concrete piles will likely drive harder, being displacement piles, and that he will probably be required to bear the additional expense associated with splicing or cutting. The following excerpt from the California Highway Division's *Standard Specifications* illustrates the usual practice of asking the contractor to bear the burden of uncertainty in predicting driven pile length.

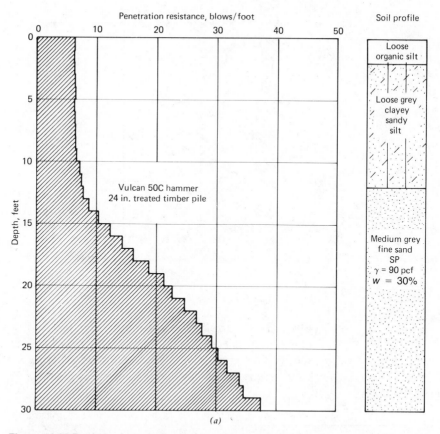

Figure 11.23 Example of penetration resistance versus depth plots for piles. *(a)* Driven in sand. "*(b)* Driven in clay. (see next page)."

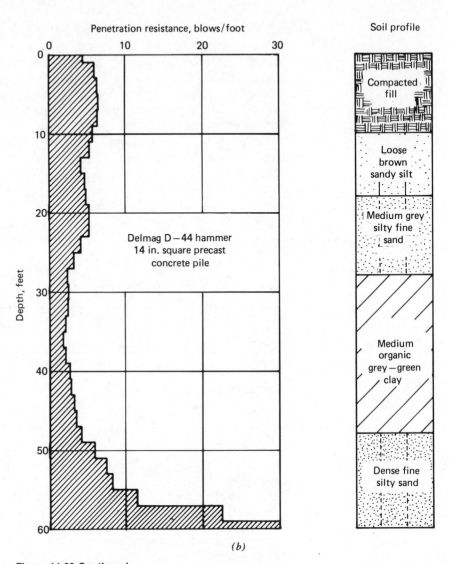

(b)

Figure 11.23 Continued

49-1.03 *Determination of Length.*
Bearing piles of any material shall be of such length as is required to develop the specified bearing value, to obtain the specified penetration, and to extend into the cap or footing block as indicated on the plans, or specified in the special provisions.

The Contractor shall be responsible for furnishing piling of sufficient length to obtain the penetration and bearing value required. For the purpose of determining the lengths of the piles required, the Contractor, at his expense, may drive test piles, make borings or make such other investigations as may be necessary. (From *Standard Specifications,* State of California, Division of Highways, 1973.)

The uncertainty is reduced if the piles are to be tip-bearing piles driven to bedrock where the bedrock surface is well defined. The uncertainty is increased if the piling are to be friction piles and if driving is to be controlled using a dynamic formula such as Equation 11.13 for capacity determinations. In preparing his bid, a contractor must account for these uncertainties. In some contracts, efforts may be made by the design engineer to eliminate some of these uncertainties and thereby obtain a better price for his client, the owner. This may be done, for instance, by specifying either final driven length or a pile order list. The latter approach is illustrated in the following excerpt from the Oregon State Highway Division's *Standard Specifications.*

503.42   Order Piling

(a) Timber, precast, and prestressed concrete piles—The Contractor shall furnish piles in accordance with an itemized order list, which shall be set forth in the special provisions, showing the number and length of all piles. The lengths given in the order list will be based on the lengths that are assumed to remain in the completed structure plus an allowance for variation in the final driven lengths. The Contractor may, at his own expense, increase the lengths given to provide for such additional length as may be necessary to suit the Contractor's method of operation.

It shall be understood that the order list may be revised by the Engineer at any time prior to or during the driving of piles with the further understanding that such change will be considered an "alteration of plans" as set forth in subsection 104.02.

(b) Steel piles and shells for cast-in-place piles—The Contractor will not be furnished an order list for steel piles or steel shells. The Contractor shall be responsible for determining the pile lengths required and for ordering and furnishing piling of sufficient length to obtain the penetration and bearing value specified, and to extend into the cap or footing as indicated on the plans. For the purpose of determining the lengths of piles required, the Contractor, at his expense, may drive test piles, make borings, or make such other investigation as he considers necessary.

The "Engineer's Estimated Length" shown on the plans or in the special provisions shall be used only for comparison of bids and for determining minimum pay lengths for driving as hereinafter specified. In no way shall the Engineer's Estimated Length be considered to represent the final individual or total pile length required. (From *Standard Specifications for Highway Construction,* Oregon State Highway Division, 1974.)

Whether or not it is important, from the contractor's viewpoint, that pile lengths are determined in advance of construction or in the field may be determined by the method of payment. There is broad latitude in the methods that have been selected. For instance, all of the piling work may be bid as a lump sum item. In this circumstance, the best bids should be made if all piling lengths are specified. Unit price bids may be more appropriate depending on the unit of measurement. Consider, for instance, the following payment provision.

49-6.02 Payment Timber, steel, and concrete piles, except cast-in-place concrete piles consisting of drilled holes filled with concrete, will be paid for at the contract price per linear foot for furnish piling and the contract unit price for drive pile. (From *Standard Specifications,* State of California, Division of Highways, 1973.)

This provision indicates that the contractor will be paid for the material furnished, regardless of the amount since the contract price for furnishing piling is based on the linear foot of measurement. The contract price for driving piling, however, is for each pile; therefore, a contractor would be paid the same amount for driving a pile to a depth of 50 ft as he would for driving to a depth of 100 ft. Alternatively, he may be paid on a price-per-driven-foot basis for driving. In either event, he should recognize the uncertainties involved and adjust his bid prices accordingly. Again, driving on a unit price for each pile driven is least risky in those situations where pile lengths are known in advance of construction. For friction piles this means that lengths must be specified. For point-bearing piles, the bedrock surface elevation must be well defined. Where pile driving is to be controlled by the application of a dynamic formula, payment to the contractor on the basis of the work actually accomplished is the most equitable. Other payment provisions such as those for splicing, cutoffs, and the use of driving aids should be clearly spelled out. Claims for extra work will be minimized if the contractor is not asked to accept the risks inherent in the uncertainties of the state of the art in pile installation that are beyond his ability to control.

11.4.6 Driving Effects

Pile-driving operations can be damaging to the piling and to structures in the surrounding area. Damage to driven piling generally results from the application of excessive hammer energy. Such damage may be manifest in broken or damaged pile tops or in a sudden change in the rate of pile penetration.

In recent years, there has been a tendency for increased damage of timber piles during installation when these piles have been designed to take large loads. In the past it has been the practice to permit loads on timber piling that range up to about 25 tons. More recently, this maximum capacity has been extended to about 50 tons. Timber piles are very often driven with control using the *Engineering News* formula. Considering that the formula has a theoretical factor of safety of 6, we can see that, while in the past the ultimate capacity of a timber piling was theoretically up to 150 tons, present practice may cause this to approach 300 tons. It is not surprising, therefore, that when this capacity is actually approached, we see an increased frequency of breakage resulting from driving.

Concrete piles may be damaged by hard driving, but also may be damaged in very soft driving conditions. When very little tip resistance is encountered and a high-capacity hammer is being used, high tensile stresses may be produced in a pile. If these tensile stresses exceed the sum of the prestressing force in a concrete pile and the tensile strength of the concrete, transverse tension cracks can be produced. These cracks can be observed to open and close with each hammer blow. Tension cracking is predictable, using the wave analysis method, and can be overcome by adequate cushioning during driving or the use of reduced hammer energy until hard driving is encountered. A sample prediction using wave analysis is illustrated in Figure 11.24.

Pile driving produces severe vibrations of the surrounding ground. The effects of this vibration may be damaging to adjacent structures. It has been

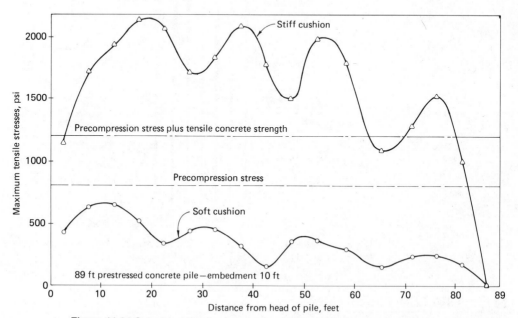

Figure 11.24 Calculated tensile stresses in a concrete pile during driving. Cracking of the pile should occur where computed tensile stress exceeds the sum of the stress resulting from prestressing and the concrete tensile strength. (After Barrero-Quiroga.)

suggested that the vibrations will be within tolerable limits at a distance of about three pile lengths from the driving operation when driving in clay soils. In sandy soils, this limit is reduced to one pile length. These are approximate relationships only and, in sensitive situations, it would be prudent to conduct special studies for potential damage resulting from driving. Pile-driving vibrations can also produce volume changes in soils. In cohesionless soils, the effect of vibrations is to produce densification, resulting in settlement of the surface of the surrounding area. Clay soils do not compact as a result of pile driving. On the contrary, when displacement piles are driven, the volume occupied by the driven piles is reflected in heave either laterally or vertically, depending on which way is easiest for the soil to move. One consequence of heave may be to raise adjacent piles that have been previously driven. For this reason, driving contracts will usually contain some provision for ascertaining the amount of this heave and for requiring subsequent redriving if it exceeds a specified limit. Lateral displacement of a clay soil during driving in one instance produced large displacements in a retaining structure that had been previously constructed. This case is illustrated in Figure 11.25.

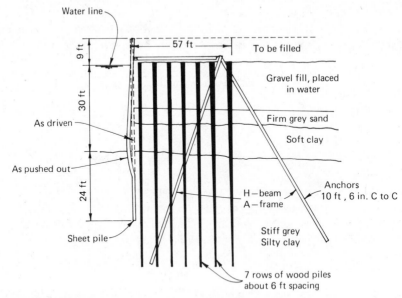

Figure 11.25 Lateral heave of sheet pile bulkhead resulting from displacement of clay soil by pile driving. (After Sowers and Sowers, 1967.)

11.5 DRILLED PIERS

Drilled piers are sometimes referred to as caissons or cast-in-place concrete piles, both because of local custom and because of the similarities between methods of construction and between the resulting foundations. The true caisson method of construction was discussed in Chapter 9 and is usually employed for larger-diameter structures and structures that include usable constructed space within the caisson, below ground elevation. Cast-in-place piles are usually of comparatively small diameter and formed by other methods. Drilled piers, then, are those deep foundations consisting of a concrete shaft from about 6 in. up to 10 ft or more in diameter, which may be equipped with an expanded tip, cast in a machine-drilled hole. Examples are shown in Figure 11.26.

11.5.1 Applications

Piers are constructed to transmit structure load to a suitable bearing stratum. One obvious application is when the surface soils are compressible, and where a suitable bearing stratum exists at reasonable depth. Another widespread application is in the control of the effects of expansive soils, where piers are used to bypass the zone of seasonal moisture change, and arrangements are made to provide a void space between the grade beam and ground surface. They are an alternative to foundation piling in some cases and may be selected over piling for a number of reasons. For instance, by using piers, it may be possible to provide very high-capacity foundations

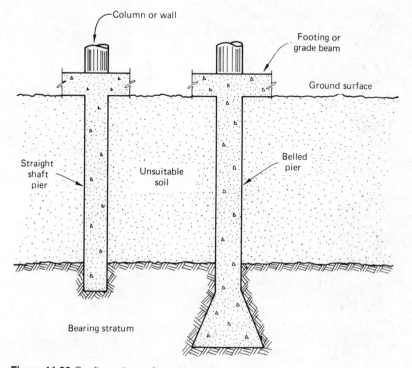

Figure 11.26 Configurations of machine-drilled piers.

without the heave or vibration problems associated with heavy pile driving. It may also be possible to establish very economical foundations of comparatively low capacity. Usually, if geologic conditions common to a geographical area are favorable for construction, piers will be a very popular foundation type. On the other hand certain geologic conditions make pier construction difficult and may favor the installation of driven piles.

11.5.2 Construction

Installing a pier involves excavation of the shaft and bell (drilling and underreaming), setting casing when required, setting reinforcing steel, placing concrete, and removing casing. Each of these operations must be done with care to produce a continuous pier with the intended shape and concrete with the specified properties.

Augers for pier drilling include both the flight and bucket types. Examples are shown in Figure 11.27. For large, deep piers the augers are mounted on a square shaft or *kelly bar* which is driven by a ring gear. On the most modern equipment the kelly is equipped with a device to provide downward

Figure 11.27 Auger types for pier drilling. *(a)* Flight augers. *(b)* A bucket auger.

force and speed drilling, which can telescope for deep work. The boring is advanced by filling the flights or bucket of the auger, raising it to the surface to discharge the soil, and reentering the hole to repeat the operation. In strongly cohesive materials an open hole may be drilled. In very soft soils or cohesionless soils it may be necessary to case the hole or drill through a clay slurry.

If a belled end is to be formed, a special bucket, such as that shown in Figure 11.28, is substituted for the auger. The belling bucket underreams the hole as its expandable sides extend. The tip diameter formed in this manner is about three times that of the main shaft.

Depending on foundation conditions, a casing may be set in the completed hole to keep it open and aid in cleaning the bottom. In some cases the casing is equipped with a cutting edge so that it may be seated in the bearing stratum. In a cased hole where a seal can be effected at the tip, and in a stable, open hole, the end of the pier can be seated on material that has been thoroughly cleaned and visually inspected.

When the hole is ready for construction of the pier itself, reinforcing is placed first if any is to be used. The steel reinforcement is usually arranged in a circular pattern of vertical bars attached to a spiral cage. All elements are connected firmly so that the reinforcement may be positioned accurately and maintained in position in the hole during the remainder of the construction operation.

In an open hole, concrete is placed down the center of the hole so that it does not impact the steel cage. In a hole filled with slurry or water, concrete is placed by the tremie method. The casing, if one is used, may be withdrawn

Figure 11.28 Belling bucket for pier-drilling machine.

Figure 11.29 Frame for withdrawing pier casings.

or left in place. If it is withdrawn, the level of the fresh concrete inside must be very carefully controlled, and the casing must be kept vertical at all times. A frame for pulling the casing for piers is shown in Figure 11.29.

11.5.3 Problems in Construction

During drilling of comparatively large-diameter piers in soft ground, unless casing or fluid is used, there is a possibility of loss of ground by squeezing toward the hole and subsequent settlement of the surrounding ground surface. If the water level in a pier hole is lower than in the surrounding area, either because of pumping or simply due to a lag during drilling, cohesionless materials may be carried to the hole with similar results. The latter situation is illustrated in Figure 11.30a and is analogous to the problems associated with sumping in cohesionless soils (see Chapter 9). Loss of ground may be prevented by casing and by maintaining the fluid level in the hole above the surrounding groundwater level at all times.

Unless the stratum in which a belled end is to be formed is strongly cohesive, the excavation may not support itself and may cave, as shown in Figure 11.30b. If this problem is not discovered by subsequent inspection, and concrete for the pier is placed, the tip-bearing area is considerably reduced, with obvious consequences. Belling should therefore be attempted only within a suitable stratum, the location of which is confirmed by com-

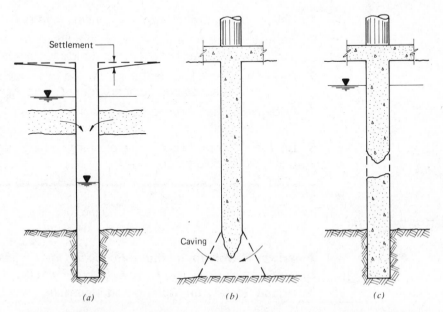

Figure 11.30 Some problems associated with drilled pier construction. *(a)* Settlement resulting from loss at ground associated with seepage toward hole. *(b)* Caving of excavated bell before concreting. *(c)* Discontinuous shaft caused by poor control during extraction of casing.

petent inspection at each pier location. No pier should ever be completed without inspection to provide assurances that the tip dimensions and conditions are as provided for in the plans.

Removal of casing is an operation that must be performed with great attention to detail in order to ensure that a situation like that in Figure 11.30c does not develop. If a casing is to be pulled, the head of concrete must always be great enough so that it will flow as the casing is withdrawn, but not so great that the flow rate is slower than the rate of casing withdrawal. Continuous inspection of the concrete surface is required to confirm that these conditions are met, and comparisons between the concrete volume and hole volume should always be made. In cases where the withdrawal of casings presents great difficulties, they should be left in place.

11.6 SUMMARY

This chapter has included a discussion of some basic concepts of foundation design and construction. The subject has been presented as an introduction to the scope of problems associated with the construction of shallow and deep foundations. Complete coverage of the state of the art in these matters would require many volumes. Consistent with the scope of this text, the reader should

1 Have learned how foundations transmit load to the soil.
2 Be able to describe the basic foundation types and their advantages and disadvantages, both from the standpoint of design and construction.
3 Have obtained a knowledge of how the basic types of foundations are built and the special equipment required for construction.
4 Understand the fundamentals of analysis of foundations for bearing capacity and settlement.
5 Be aware of some of the problems arising during foundation construction, and how to solve them.

REFERENCES

American Petroleum Institute, *Recommended Practice for Planning, Designing, and Constructing Fixed Offshore Platforms,* April 1977.

American Society for Testing and Materials, *1974 Book of ASTM Standards,* Philadelphia.

Barrero-Quiroga, Franklin, *Pile Hammer Selection by Driving Stress Analysis,* Unpublished M.S. Thesis, Oregon State University, 1974.

Bethlehem Steel Corporation, *Steel H-Piles,* Bethlehem, Pa.

Chellis, R. D., *Pile Foundations,* McGraw-Hill, New York, 1961.

Hagerty, D. J., and Peck, R. B., "Heave and Lateral Movements due to Pile Driving," *Journal, Soil Mechanics and Foundations Division, American Society of Civil Engineers,* **97,** No. SM11, 1971.

Hirsch, T. J., Lowery, L. L., Coyle, H. M., and Samson, C. H., "Pile-driving Analysis by One Dimensional Wave Theory: State of the Art," *Highway Research Record* 333, 1970.

Luna, William A., "Ground Vibrations due to Pile Driving," *Foundation Facts, Raymond International,* **3,** No. 2, 1967.

Mayo, David F., "Development of Nonpneumatic Caisson Engineering," *Journal, Construction Division, American Society of Civil Engineers,* **99,** No. C01, 1973.

Meyerhof, G. G., "Shallow Foundations," *Journal, Soil Mechanics and Foundations Division, American Society of Civil Engineers,* **91,** No. SM2, 1965.

Oregon State Highway Division, *Standard Specifications for Highway Construction,* 1974.

Peck, R. B., Hanson, W. E., and Thornburn, T. H., *Foundation Engineering,* Wiley, New York, 1974.

Peurifoy, R. L., *Construction Planning, Equipment, and Methods,* McGraw-Hill, New York, 1970.

Romanoff, M., *Corrosion of Steel Pilings in Soils,* United States Bureau of Commerce, National Bureau of Standards, Monograph 58, 1962.

Smith, E. A. L., "Pile-driving Analysis by the Wave Equation," *Journal, Soil Mechanics and Foundations Division, American Society of Civil Engineers,* **86,** No. SM4, 1960.

Sowers, G. B., and Sowers, G. F., "Failures of Bulkhead and Excavation Bracing," *Civil Engineering,* January 1967.

State of California, Division of Highways, *Standard Specifications,* 1973.

Terzaghi, K., and Peck, R. B., *Soil Mechanics in Engineering Practice,* Wiley, New York, 1967.

U.S. Bureau of Reclamation, *Concrete Manual,* 1966.

PROBLEMS

1 Footings for a structure are to be built beneath a basement slab on a clay soil at a depth of 20 ft below present ground level. Representative consolidation tests on the soil produce results typified by Figure 5.9. The groundwater table is 15 ft below the ground surface. The soil-void ratio is about 0.87. If the average bearing pressure on the footings is 2600 psf, estimate the settlement of 10-ft and 5-ft-square footings.

2   What is the factor of safety for the footings in Problem 1 against bearing failure if the clay soil has an unconfined compressive strength of 3600 psf?

3   What is the maximum column load in kips that could be placed on a 4-ft square temporary footing on a sand for which the standard penetration resistance is about $N = 15$ and the water table is at the surface? A factor of safety of 2 is to be maintained against bearing failure.

4   Solve Problem 3 if settlement of the footing must be limited to ½ in.

5   Six hundred timber piles are to be driven to a depth of 40 ft in loose sand and silt below the water table. Compute a representative ultimate pile capacity if minimum tip diameter is 8 in. and minimum butt diameter is 20 in. Suggest a type and size of pile-driving hammer. The pile design load is 25 tons. Justify your choice and describe what problems you might encounter.

6   Prestressed concrete friction piles, 12 in. square, are to be driven to a working load of 40 tons in a soft to medium clay deposit according to the provisions of a contract you are preparing a bid for. Capacity is to be determined by the *Engineering News* formula, and lengths are not specified. Select a tentative hammer type and size, justifying your choice. How long should the piles you furnish be? Describe problems you may encounter.

7   Steel H-piles are to be driven to tip bearing through about 150 ft of soft clay and plastic silt. The design capacity of the piles is 100 tons. What type of hammer would you use? What special problems might you encounter that should be considered in bidding the work?

8   Specifications require that timber piles be driven to a depth of 50 ft in a stiff clay soil. Minimum tip diameter is 8 in. and minimum butt diameter is 20 in. The specifications further require a minimum demonstrated allowable capacity of 40 tons, with a factor of safety of at least 3. Calculated capacities from driving records for the first few piles according to the *Engineering News* formula are less than 25 tons. You have all 55-ft piles on hand. How would you proceed? Justify your answer.

9   Figure 10.15 shows the soil profile at a building site. Pier foundations are to be used and design requirements dictate two alternatives. The first employs belled piers with 6-ft-diameter tips at 26 ft in depth. The second requires 2-ft-diameter straight shafts to 42 ft in depth. Which alternative would you prefer to build? Justify your answer, giving the advantages and disadvantages of each alternative.

CHAPTER 12

CONSTRUCTION ACCESS AND HAUL ROADS

On earth-moving projects and many other jobs, temporary roads are built to move materials to the project site. In some cases, existing roads are used, and often they are upgraded for this purpose. Sometimes such roads become part of the permanent works. In other cases, haul roads are temporary only. Public roads are sometimes used for access and occasionally must accommodate very heavy traffic and loads. Access and haul roads may be surfaced or unsurfaced, according to project requirements. The purposes of this chapter are to

1 Give the reader an introduction to pavement and subgrade structural performance requirements.
2 Show how subgrade strength and bearing capacity analysis may be employed to design or evaluate construction roads.
3 Provide an introduction to the rationale for commonly employed road maintenance and improvement procedures.

12.1 PAVEMENT COMPONENTS AND THEIR FUNCTIONS

Figure 12.1 shows a cross section through a pavement. The pavement consists of the higher quality (usually imported) materials above the subgrade, including the wearing surface, the base course, and the subbase. The function of these materials is to protect the subgrade from traffic loads and weather. Their quality and thickness requirements will be determined by

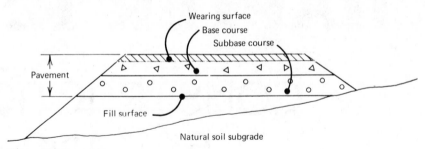

Figure 12.1 Pavement section.

subgrade conditions and traffic loadings. In many instances, particularly depending on subgrade conditions and the weather, some construction roads may be built to be used without a pavement.

The wearing surface on any road is the highest quality material in the pavement structure, except in those instances when the subgrade consists of hard, durable intact rock. On permanent roads it usually consists of asphalt concrete, Portland cement concrete, or stabilized aggregate. It must be sufficiently strong to resist compressive stresses from wheel loads and cohesive enough to withstand traffic abrasion.

The base course is usually the major structural element in a pavement, except for those surfaced with Portland cement concrete. It consists of a comparatively thick, highly stable layer of well-graded clean, coarse, crushed aggregate. Uncrushed aggregate is sometimes used when suitable natural gradations are available. Subbases are employed in permanent works but not often in construction pavements. The quality of the subbase is intermediate—between base course and subgrade.

12.2 SUBGRADE MATERIALS AND STRENGTH REQUIREMENTS

The general suitability of various soils as subgrade materials may be inferred from soil classifications. Such correlations are shown in Figure 4.12 for the AASHTO system and in Figure 4.15 for the Unified System. Beyond general suitability, subgrade strength is evaluated by methods used in conjunction with various pavement design procedures. Most of the methods used for pavement design are empirical or semiempirical. Each requires complex equipment and tests to determine subgrade strength, and each employs very specialized analytical procedures. These pavement design methods are beyond the scope of this text. A reasonably good understanding of pavement analysis methods and subgrade material strength requirements can be acquired, however, by considering the general bearing capacity and stress analysis methods discussed in Section 11.2.1, and by applying these theoretical procedures to the particular problem of designing a pave-

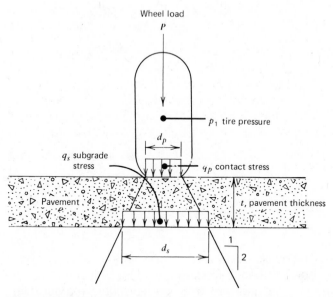

Figure 12.2 Definition of terms describing wheel load transfer through pavement to subgrade.

ment to support a loaded wheel. Figure 12.2 provides a summary of the necessary definitions. Table 12.1 gives some examples of wheel loads for typical construction vehicles.

For usual wheel loads a circular contact "print" of the tire is presumed. The contact stress q_p is equal to the tire pressure p. The diameter of the "print" or loaded area is thus derived from

$$P = q_p \frac{\pi}{4} d_p^2 = p \frac{\pi}{4} d_p^2 \qquad (12.1)$$

TABLE 12.1   TYPICAL WHEEL LOADS FOR CONSTRUCTION VEHICLES

Vehicle	Maximum Axle Load (kips)	Maximum Wheel Load (kips)	Tire Pressure[a] (psi)
Highway truck, legal, single axle, dual tires	18–20	9–10	60–90
Highway truck, maximum, single axle, dual tires	32	16	60–90
Front loader, light	29	15	50–90
Front loader, heavy	118	59	50–90
Off highway truck, light	66	33	50–90
Off highway truck, heavy	86	43	50–90
Scraper, light	30	15	50–90
Scraper, medium	57	28	50–90
Scraper, heavy	94	47	50–90

[a] Tire pressures are varied to optimize performance for given loadings and road conditions.

as

$$d_p = \sqrt{\frac{4P}{p\pi}} \tag{12.2}$$

For multiple-wheel configurations, loads may be superimposed to calculate a pressure distribution and "print" on the subgrade surface, using the method described in Section 11.11.1. If we now assume that stresses from a single wheel are distributed beneath the loaded area in Figure 12.2 over the width

$$d_s = (d_p + t) \tag{12.3}$$

then, from Equation 12.2, the width of the loaded area on the subgrade is

$$d_s = \sqrt{\frac{4P}{p\pi}} + t \tag{12.4}$$

Equation 11.4 relates the ultimate bearing capacity of the soil beneath a loaded area to the strength of the soil and the size of the area. If, in Figure 11.3, the area above the loaded surface *a-a* were filled with the same material as exists outside the width *B*, the net ultimate bearing capacity of the subgrade, from Equation 11.4, would be

$$q_d - \gamma D_f = \frac{\gamma B}{2} N_\gamma + \gamma D_f N_q + 1.2cN_c - \gamma D_f$$

or

$$q_{d(net)} = \frac{\gamma B}{2} N_\gamma + \gamma D_f(N_q - 1) + 1.2cN_c \tag{12.5}$$

For the terminology shown in Figure 12.2, from Equation 12.5 we find that the ultimate strength of the subgrade q_{sd} will be

$$q_{sd} = \frac{\gamma d_s}{2} N_\gamma + \gamma t(N_q - 1) + 1.2cN_c \tag{12.6}$$

When this strength is compared to the subgrade stress q_s, we have an expression for the factor of safety against bearing failure beneath the pavement:

$$\frac{q_{sd}}{q_s} = FS \tag{12.7}$$

Normally, the value of the factor of safety should exceed about 2, to prevent plastic deformation in the subgrade.

To illustrate the use of Equation 12.7, suppose we wish to know whether a pavement consisting of 12 in. of crushed rock over a medium ($c = 750$ psf) clay subgrade is adequate to support a 10,000-lb single-wheel load. Tire pressure is 80 psi.

First, from Equation 12.2,

$$d_p = \sqrt{\frac{4P}{p\pi}} = \sqrt{\frac{4 \times 10,000}{80 \times \pi}} = 12.62 \text{ in.}$$

Also, from Equation 12.3,

$$d_s = d_p + t = 12.62 + 12 = 24.62 \text{ in.}$$

and the subgrade stress resulting from the wheel load is found by dividing the load by the area over which the subgrade stress is applied.

$$q_s = \frac{4P}{\pi d_s^2} = \frac{4 \times 10,000}{\pi \times 24.62^2} = 21.0 \text{ psi} = 3025 \text{ psf}$$

Now, from Equation 12.6, for the $\phi = 0$ case,

$$q_{sd} = 1.2cN_c = 6.2c = 6.2 \times 750 = 4650 \text{ psf}$$

From Equation 12.7,

$$FS = \frac{q_{sd}}{q_s} = \frac{4650}{3025} = 1.54$$

The result of the analysis indicates that the rock thickness is not sufficient to "protect" the subgrade against excessive stresses with the usual factor of safety.

The foregoing procedure considers only a single application of the wheel load. More widely used procedures consider the effects of numbers of load applications on pavement performance. Thus, it is possible that one can utilize a "pavement life" concept in arriving at a design, by evaluating the expected amount of traffic and the magnitude of the wheel loads.

12.3 SUBGRADE IMPROVEMENT

The previous section displayed how theoretical methods may be used to evaluate the suitability of a pavement for a single passage of a wheel load. Should analysis by these or other pavement analysis methods show that the pavement is not suitable, two options are available to us. The first is to use a thicker pavement and the second is to improve or strengthen the subgrade. As indicated in Chapter 8, subgrade improvement can be accomplished by compaction or by chemical stabilization. These procedures would of course be applicable only to newly built roadways. For example, it has been shown that for cohesionless sands the angle of internal friction can be related approximately to relative density by

$$\phi' = 25° + 0.15D_r \qquad (12.8)$$

Thus, by going from $D_r = 30$ percent to $D_r = 80$ percent as a result of compaction, we obtain an angle of internal friction increase from about 30° to about 37°. Referring to Figure 11.4, one can see that the bearing capacity factors N_γ and N_q to be applied in Equation 12.6 are more than doubled as a result. Obviously, compaction has a beneficial effect on subgrade strength.

Chapter 8 also provides some indication of how the cohesive strength of some fine soils may be increased by mixing with chemicals and how cohesion may be developed in sands and gravels by the addition of cement. For both soil types, in real applications, studies are usually conducted with on-site soils and proposed stabilizing agents to determine what strength improvements might be available.

Laboratory studies have shown that the relationships in Figure 12.3 represent the 28-day unconfined compressive strengths to be expected when

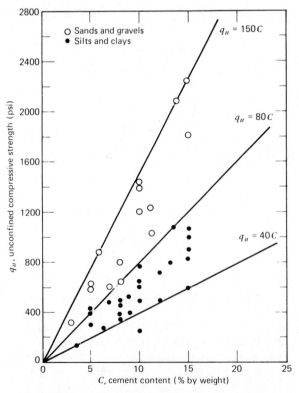

Figure 12.3 Relationships between cement content and 28-day unconfined compressive strength for soil and cement mixtures. (Adapted from FHWA-1P-80-2, October 1979.)

mixing cement with various types of soils. Mixtures of soil and cement will continue to gain strength over an extended period, and the gain would be expected to be approximated by the relation

$$q_{ud} = q_{ud_0} + K \log_{10} (d/d_0) \qquad (12.9)$$

where

q_{ud} = unconfined compressive strength at an age of d days, in psi
q_{ud_0} = unconfined compressive strength at an age of d_0 days, in psi
K = $70C$ for granular soils and $10C$ for fine-grained soils
C = cement content, in percent by weight

So for a silty soil with a 6 percent cement content which has a 7-day strength of 200 psi, we would expect that the strength at 60 days would be

$$q_{u60} = 200 + 10(6) \log_{10} (60/7) = 256 \text{ psi}$$

The cohesive strength to be used in Equation 12.6 is, of course, by definition one-half the unconfined compressive strength.

For lime treatment of road subgrades selected laboratory and field results are presented in Table 12.2. Note that, in general, the field strength values shown are greater than the laboratory values. This is not usually the case since there is less thorough mixing in the field operation. The field data, in this case, however, represents subgrades in place for considerable time periods. In practice, field strengths are usually taken as equal to laboratory values, and extra lime is added to offset mixing problems. No known simplified relationships for the effect of lime content on soil strengths are known.

Recently, a number of geotextiles, or engineering fabrics, have been developed, and these fabrics have a number of useful applications in the construction field. One of these applications is strengthening roads on weak fine soils. In practice, geotextiles may be laid directly on the subgrade and covered with an aggregate surfacing. In general, for geotextile reinforced roads on fine soils, Equation 12.6 reduces to

$$q_{sd} = 1.2cN_c = 6.2c \qquad (12.10)$$

and, because the presence of the fabric restricts plastic deformation, may be applied with a factor of safety as low as 1. Various methods are available for analyzing the degree of improvement in specific cases, and it has been shown that all available fabrics do not provide the same beneficial effects. Road reinforcement using fabrics on soft subgrades has, however, been proven in practice as a means of reducing the pavement thickness required and it is worthy of consideration where significant savings may be realized.

12.4 MAINTENANCE REQUIREMENTS

During a project it may be necessary to restore a road to its original condition to improve its riding qualities. When a project is finished, it is sometimes required that the condition of preexisting roads be reestablished.

TABLE 12.2 UNCONFINED COMPRESSIVE STRENGTH SUMMARY

Test Site	Classification	Age (yr)	Lime (%)	Unstabilized Subgrade		Lime Stabilized Subgrade							
						Field Testing Program				Laboratory Testing Program			
				q_u (psi)	Average q_u	w (%)	γ_d (pcf)	q_u (psi)	Average q_u	w (%)	γ_d (pcf)	q_u (psi)	Average q_u
LEAN CLAY SUBGRADES													
Perry Co., Missouri Location #1/Test Site 2-66	A-7-6(18)/CL	14	4.5	36 21	29					21.1 15.4	105.4 105.8	86 51	68
Ft. Hood, Kileen, Texas Test Site 4-66	A-5(11)/CL	4	4.0	28 35 27 26 41 33	32					13.2 12.7	111.9 110.6	133 152	142
Frederick Co., Virginia Test Site 7-67	A-6(16)/CL	5	7.0			16.2	109.5	267	267	16.5 15.9 15.4	111.2 106.5 107.4	61 60 64	62
FAT CLAY SUBGRADES													
Bergstrom AFB Austin, Texas	A-7-5(19)/CH	9	4.0	7 13	10					18.0 17.7	97.1 102.5	66 101	83
Giles Co., Virginia Location #1/Test Site 6-67	A-7-5(18)/CH	6	7.0			27.3	89.6	171	171				
Location #2	A-7-5(18)/CH	6	7.0	22	22	20.7	99.2	268	268	21.4 21.8	105.5 92.0	116 154	135

Source: From R. E. Aufmuth, "Strength and Durability of Stabilized Layers under Existing Pavements," *Technical Report M-4, Construction Engineering Research Laboratory*, Champaign, Illinois, 1970.

For paved roads with a high-quality wearing surface, patching and repairs at damaged locations may be all that is required. Alternatively, overlays or new surfaces may be necessary. Whether for the contractor's own benefit or for post-project service, the need for such improvements should be evaluated in the estimating and bidding process and the appropriate costs included in the proposal. In making such an evaluation, one may consider the feasibility of using light hauling equipment, which will not adversely impact the road, versus the option of using heavy equipment and repairing the consequent damage.

For aggregate surfaced or unsurfaced roads, maintenance during construction usually involves watering, regrading, and compaction. Light watering is principally used to control dust. Maintenance of the subgrade at near optimum water contact, however, results in better control of rutting and deformation by improving subgrade strength. The value of maintaining an even, dense road surface can be assessed by considering the rolling resistance of equipment. Rolling resistance may be defined as the force required to move a given unit of weight over the surface in question. It is typically expressed in pounds per ton and depends not only on road surface condition, but also on equipment characteristics. Soft, deformable road surfaces have higher rolling resistance because greater energy must be exerted to cause a wheel to climb out of the depression created by the load it carries. Underinflation of tires results in higher rolling resistances on uniform, nonyielding surfaces. Some example values of rolling resistance are provided in Table 12.3.

The selection of haul roads for construction equipment, especially for long

TABLE 12.3   REPRESENTATIVE ROLLING RESISTANCES FOR VARIOUS TYPES OF WHEELS AND SURFACES (IN POUNDS PER TON OF GROSS LOAD)

Type of Surface	Steel Tires, Plain Bearings	Crawler-Type Track and Wheel	Rubber Tire, Antifriction Bearings	
			High Pressure	Low Pressure
Smooth concrete	40	55	35	45
Good asphalt	50–70	60–70	40–65	50–60
Earth, compacted and well maintained	60–100	60–80	40–70	50–70
Earth, poorly maintained, rutted	100–150	80–110	100–140	70–100
Earth, rutted, muddy, no maintenance	200–250	140–180	180–220	150–200
Loose sand and gravel	280–320	160–200	260–290	220–260
Earth, very muddy, rutted, soft	350–400	200–240	300–400	280–340

Source: From R. L. Peurifoy, *Construction Planning, Equipment and Methods,* McGraw-Hill Book Company, New York, 1970.

distances or heavy loads, depends on a number of factors, including road grade and machinery available. Table 12.3 shows that just the maintenance of haul roads on large projects can be a wise investment that considerably reduces power requirements and produces significant fuel savings.

12.5 SUMMARY

Access and haul roads can be a significant consideration on many construction projects. Requirements for road construction, improvement, and maintenance are affected by site conditions, weather, equipment type, and expected traffic. Pavement design and analysis procedures can be used to evaluate the different factors of a particular job site. The reader should

1 Understand terminology associated with pavement components.
2 Be able to make a theoretical evaluation of the suitability of a given pavement for a given wheel load.
3 Be aware that more detailed analyses are required to account for effects of nonhomogeneous paving materials and repeated traffic loads.
4 Appreciate that substantial improvements in subgrade behavior may be achieved by chemical stabilization, compaction, or the use of geotextiles.
5 Recognize the importance of a planned maintenance program for haul roads.

REFERENCES

American Association of State Highway Officials, *Standard Specifications for Highway Bridges,* Washington, D.C., 1965.

Aufmuth, R. E., "Strength and Durability of Stabilized Layers under Existing Pavements," *Technical Report M-4, Construction Engineering Research Laboratory,* Champaign, Illinois, 1970.

Giroud, J. P., and Noiray, L., "Geotextile-Reinforced Unpaved Road Design," *Journal, Geotechnical Engineering Division, American Society of Civil Engineers,* **107,** No. GT9, 1981.

Peurifoy, R. L., *Construction Planning, Equipment and Methods,* McGraw-Hill, New York, 1970.

U.S. Department of Transportation, *Soil Stabilization in Pavement Structures—A User's Manual, Vol. 2,* FHWA-1P-80-2, 1979.

Yoder, E. J., *Principles of Pavement Design,* Wiley, New York, 1959.

PROBLEMS

1   A front loader will operate in an aggregate crushing operation on river bottom land. The soil is saturated with an unconfined compressive strength averaging about 900 psf. The maximum wheel load for the loader proposed is 40 kips. Tire pressure is 70 psi. Calculate the thickness of aggregate that should be placed over the natural soil to provide a suitable working surface.

2   Off-highway trucks with maximum 80-kip axle loads and 60-psi tire pressures will operate on a sand fill. The single rear axle has dual tires 2 ft apart. Compare the suitability of the fill to support the trucks after it has been compacted ($D_r = 85$ percent) to that after it was placed ($D_r = 55$ percent).

3   A private farm road is to be used for hauling aggregate from a borrow pit to a construction project. The road is constructed with a 4-in. asphalt stabilized aggregate surface over a clay subgrade with an unconfined compressive strength of about 3000 psf. Legal highway dump trucks will be used. The farmer has stipulated that the road must be restored to its original condition after construction. Is it likely that significant repairs will be required?

4   A 5-mile-long haul road over sandy soils will be used to haul borrow to a dam site. Large off-highway trucks and scrapers will be used. The natural relative density of the sand is about 50 percent. Estimate the thickness of cement-stabilized sand using 8 percent cement that would provide a pavement giving strength equivalent to a 2-foot-thick compacted zone at $D_r = 75$ percent.

5   Compare the estimated thicknesses of gravel required to support an overloaded highway truck (32-kip axle load) on a soft subgrade ($q_u = 750$ psf) with and without a geotextile at the fabric-soil interface.

PART THREE

APPENDIX OF TESTING METHODS*

Standard Method for

DRY PREPARATION OF SOIL SAMPLES FOR PARTICLE-SIZE ANALYSIS AND DETERMINATION OF SOIL CONSTANTS[1]

This standard is issued under the fixed designation D 421; the number immediately following the designation indicates the year of original adoption or, in the case of revision, the year of last revision. A number in parentheses indicates the year of last reapproval.

1. Scope

1.1 This method covers the dry preparation of soil samples as received from the field for particle-size analysis and the determination of the soil constants.

2. Apparatus

2.1 *Balance*—A balance sensitive to 0.1 g.

2.2 *Mortar*—A mortar and rubber-covered pestle suitable for breaking up the aggregations of soil particles.

2.3 *Sieves*—A series of sieves, of square mesh woven wire cloth, conforming to ASTM Specification E 11, for Wire-Cloth Sieves for Testing Purposes.[2] The sieves required are as follows:

    No. 4 (4.75-mm)
    No. 10 (2.00-mm)
    No. 40 (425-μm)

2.4 *Sampler*—A riffle sampler or sample splitter, for quartering the samples.

3. Sampling

3.1 Expose the soil sample as received from the field to the air at room temperature until dried thoroughly. Break up the aggregations thoroughly in the mortar with a rubber-covered pestle. Select a representative sample of the amount required to perform the desired tests by the method of quartering or by the use of a sampler. The amounts of material required to perform the individual tests are as follows:

3.1.1 *Particle-Size Analysis*—For the particle-size analysis, material passing a No. 10 (2.00-mm) sieve is required in amounts equal to 115 g of sandy soils and 65 g of either silt or clay soils.

3.1.2 *Tests for Soil Constants*—For the tests for soil constants, material passing the No. 40 (425-μm) sieve is required in total amount of 220 g, allocated as follows:

Test	Grams
Liquid limit	100
Plastic limit	15
Centrifuge moisture equivalent	10
Volumetric shrinkage	30
Check tests	65

4. Preparation of Test Sample

4.1 Select that portion of the air-dried sample selected for purpose of tests and record the mass as the mass of the total test sample uncorrected for hygroscopic moisture. Separate the test sample by sieving with a No. 10 (2.00-mm) sieve. Grind that fraction retained on the No. 10 sieve in a mortar with a rubber-covered pestle until the aggregations of soil particles are broken up into the separate grains. Then separate the ground soil into two fractions by sieving with a No. 10 sieve.

4.2 Wash that fraction retained after the second sieving free of all fine material, dry, and weigh. Record this mass as the mass of coarse material. Sieve the coarse material after being washed and dried, on the No. 4 (4.75-mm) sieve and record the mass retained on the No. 4 sieve.

This method is under the jurisdiction of ASTM Committee D-18 on Soil and Rock.
Current edition approved Sept. 22, 1958. Originally issued 1935. Replaces D 421 – 38.
[2] *Annual Book of ASTM Standards*, Part 41.

D 421

5. Test Sample for Particle-Size Analysis

5.1 Mix the fractions passing the No. 10 (2.00-mm) sieve in both sieving operations thoroughly together, and by the method of quartering or the use of a sampler, select a portion weighing approximately 115 g for sandy soils and approximately 65 g for silt and clay soil for particle-size analysis.

6. Test Sample for Soil Constants

6.1 Separate the remaining portion of the material passing the No. 10 (2.00-mm) sieve into two parts by means of a No. 40 (425-μm) sieve. Discard the fraction retained on the No. 40 sieve. Use the fraction passing the No. 40 sieve for the determination of the soil constants.

ASTM D 422 – 63 (Reapproved 1972)

Standard Method for
PARTICLE-SIZE ANALYSIS OF SOILS[1]

This standard is issued under the fixed designation D 422; the number immediately following the designation indicates the year of original adoption or, in the case of revision, the year of last revision. A number in parentheses indicates the year of last reapproval.

1. Scope

1.1 This method covers the quantitative determination of the distribution of particle sizes in soils. The distribution of particle sizes larger than 75 μm (retained on the No. 200 sieve) is determined by sieving, while the distribution of particle sizes smaller than 75 μm is determined by a sedimentation process, using a hydrometer to secure the necessary data (Notes 1 and 2).

NOTE 1—Separation may be made on the No. 4 (4.75-mm), No. 40 (425-μm), or No. 200 (75-μm) sieve instead of the No. 10. For whatever sieve used, the size shall be indicated in the report.

NOTE 2—Two types of dispersion devices are provided: (1) a high-speed mechanical stirrer, and (2) air dispersion. Extensive investigations indicate that air-dispersion devices produce a more positive dispersion of plastic soils below the 20-μm size and appreciably less degradation on all sizes when used with sandy soils. Because of the definite advantages favoring air dispersion, its use is recommended. The results from the two types of devices differ in magnitude, depending upon soil type, leading to marked differences in particle size distribution, especially for sizes finer than 20 μm.

2. Apparatus

2.1 *Balances*—A balance sensitive to 0.01 g for weighing the material passing a No. 10 (2.00-mm) sieve, and a balance sensitive to 0.1 percent of the mass of the sample to be weighed for weighing the material retained on a No. 10 sieve.

2.2 *Stirring Apparatus*—Either apparatus A or B may be used.

2.2.1 Apparatus A shall consist of a mechanically operated stirring device in which a suitably mounted electric motor turns a vertical shaft at a speed of not less than 10,000 rpm without load. The shaft shall be equipped with a replaceable stirring paddle made of metal, plastic, or hard rubber, as shown in Fig. 1. The shaft shall be of such length that the stirring paddle will operate not less than ³⁄₄ in. (19.0 mm) nor more than 1¹⁄₂ in. (38.1 mm) above the bottom of the dispersion cup. A special dispersion cup conforming to either of the designs shown in Fig. 2 shall be provided to hold the sample while it is being dispersed.

2.2.2 Apparatus B shall consist of an air-jet dispersion cup[2] (Note 3) conforming to the general details shown in Fig. 3 (Notes 4 and 5).

NOTE 3—The amount of air required by an air-jet dispersion cup is of the order of 2 ft³/min; some small air compressors are not capable of supplying sufficient air to operate a cup.

NOTE 4—Another air-type dispersion device, known as a dispersion tube, developed by Chu and Davidson at Iowa State College, has been shown to give results equivalent to those secured by the air-jet dispersion cups. When it is used, soaking of the sample can be done in the sedimentation cylinder, thus eliminating the need for transferring the slurry. When the air-dispersion tube is used, it shall be so indicated in the report.

NOTE 5—Water may condense in air lines when not in use. This water must be removed, either by using a water trap on the air line, or by blowing the water out of the line before using any of the air for dispersion purposes.

2.3 *Hydrometer*—An ASTM hydrometer, graduated to read in either specific gravity of the suspension or grams per litre of suspension, and conforming to the requirements for

---

[1] This method is under the jurisdiction of ASTM Committee D-18 on Soil and Rock.

Current edition accepted Nov. 21, 1963. Originally issued 1935. Replaces D 422 – 62.

[2] Detailed working drawings for this cup are available at a nominal cost from the American Society for Testing and Materials, 1916 Race St., Philadelphia, Pa. 19103. Order Adjunct No. 12-404220-00.

ASTM **D 422**

hydrometers 151H or 152H in ASTM Specification E 100, for ASTM Hydrometers.[3] Dimensions of both hydrometers are the same, the scale being the only item of difference.

2.4 *Sedimentation Cylinder*—A glass cylinder essentially 18 in. (457 mm) in height and 2½ in. (63.5 mm) in diameter, and marked for a volume of 1000 ml. The inside diameter shall be such that the 1000-ml mark is 36 ± 2 cm from the bottom on the inside.

2.5 *Thermometer*—A thermometer accurate to 1 F (0.5 C).

2.6 *Sieves*—A series of sieves, of square-mesh woven-wire cloth, conforming to the requirements of ASTM Specification E 11, for Wire-Cloth Sieves for Testing Purposes.[3] A full set of sieves includes the following (Note 6):

3-in. (75-mm)	No. 10 (2.00-mm)
2-in. (50-mm)	No. 20 (850-µm)
1½-in. (37.5-mm)	No. 40 (425-µm)
1-in. (25.0-mm)	No. 60 (250-µm)
¾-in. (19.0-mm)	No. 140 (106-µm)
⅜-in. (9.5-mm)	No. 200 (75-µm)
No. 4 (4.75-mm)	

NOTE 6—A set of sieves giving uniform spacing of points for the graph, as required in Section 16, may be used if desired. This set consists of the following sieves:

3-in. (75-mm)	No. 16 (1.18-mm)
1½-in. (37.5-mm)	No. 30 (600-µm)
¾-in. (19.0-mm)	No. 50 (300-µm)
⅜-in. (9.5-mm)	No. 100 (150-µm)
No. 4 (4.75-mm)	No. 200 (75-µm)
No. 8 (2.36-mm)	

2.7 *Water Bath or Constant-Temperature Room*—A water bath or constant-temperature room for maintaining the soil suspension at a constant temperature during the hydrometer analysis. A satisfactory water tank is an insulated tank that maintains the temperature of the suspension at a convenient constant temperature at or near 68 F (20 C). Such a device is illustrated in Fig. 4. In cases where the work is performed in a room at an automatically controlled constant temperature, the water bath is not necessary.

2.8 *Beaker*—A beaker of 250-ml capacity.

2.9 *Timing Device*—A watch or clock with a second hand.

3. Dispersing Agent

3.1 A solution of sodium hexametaphosphate (sometimes called sodium metaphosphate) shall be used in distilled or demin-eralized water, at the rate of 40 g of sodium hexametaphosphate/litre of solution (Note 7).

NOTE 7—Solutions of this salt, if acidic, slowly revert or hydrolyze back to the orthophosphate form with a resultant decrease in dispersive action. Solutions should be prepared frequently (at least once a month) or adjusted to pH of 8 or 9 by means of sodium carbonate. Bottles containing solutions should have the date of preparation marked on them.

3.2 All water used shall be either distilled or demineralized water. The water for a hydrometer test shall be brought to the temperature that is expected to prevail during the hydrometer test. For example, if the sedimentation cylinder is to be placed in the water bath, the distilled or demineralized water to be used shall be brought to the temperature of the controlled water bath; or, if the sedimentation cylinder is used in a room with controlled temperature, the water for the test shall be at the temperature of the room. The basic temperature for the hydrometer test is 68 F (20 C). Small variations of temperature do not introduce differences that are of practical significance and do not prevent the use of corrections derived as prescribed.

4. Test Sample

4.1 Prepare the test sample for mechanical analysis as outlined in ASTM Method D 421. Dry Preparation of Soil Samples for Particle-Size Analysis and Determination of Soil Constants.[4] During the preparation procedure the sample is divided into two portions. One portion contains only particles retained on the No. 10 (2.00-mm) sieve while the other portion contains only particles passing the No. 10 sieve. The mass of air-dried soil selected for purpose of tests, as prescribed in Method D 421, shall be sufficient to yield quantities for mechanical analysis as follows:

4.1.1 The size of the portion retained on the No. 10 sieve shall depend on the maximum size of particle, according to the following schedule:

Nominal Diameter of Largest Particles, in. (mm)	Approximate Minimum Mass of Portion, g
⅜ (9.5)	500
¾ (19.0)	1000

[3] *Annual Book of ASTM Standards*, Part 41.
[4] *Annual Book of ASTM Standards*, Part 19

Nominal Diameter of Largest Particles, in. (mm)	Approximate Minimum Mass of Portion, g
1 (25.4)	2000
1½ (38.1)	3000
2 (50.8)	4000
3 (76.2)	5000

4.1.2 The size of the portion passing the No. 10 sieve shall be approximately 115 g for sandy soils and approximately 65 g for silt and clay soils.

4.2 Provision is made in Section 4 of Method D 421 for the weighing of the air-dry soil selected for purpose of tests, the separation of the soil on the No. 10 sieve by dry-sieving and washing, and the weighing of the washed and dried fraction retained on the No. 10 sieve. From these two masses the percentages retained and passing the No. 10 sieve can be calculated in accordance with 11.1.

NOTE 8—A check on the mass values and the thoroughness of pulverization of the clods may be secured by weighing the portion passing the No. 10 sieve and adding this value to the mass of the washed and oven-dried portion retained on the No. 10 sieve.

SIEVE ANALYSIS OF PORTION RETAINED ON NO. 10 (2.00-mm) SIEVE

5. Procedure

5.1 Separate the portion retained on the No. 10 (2.00-mm) sieve into a series of fractions using the 3-in. (75-mm), 2-in. (50-mm), 1½-in. (37.5-mm), 1-in. (25.0-mm), ¾-in. (19.0-mm), ⅜-in. (9.5-mm), No. 4 (4.75-mm), and No. 10 sieves, or as many as may be needed depending on the sample, or upon the specifications for the material under test.

5.2 Conduct the sieving operation by means of a lateral and vertical motion of the sieve, accompanied by a jarring action in order to keep the sample moving continuously over the surface of the sieve. In no case turn or manipulate fragments in the sample through the sieve by hand. Continue sieving until not more than 1 mass percent of the residue on a sieve passes that sieve during 1 min of sieving. When mechanical sieving is used, test the thoroughness of sieving by using the hand method of sieving as described above.

5.3 Determine the mass of each fraction on a balance conforming to the requirements of 2.1. At the end of weighing, the sum of the masses retained on all the sieves used should

A$\widehat{\text{S}}$TM D 422

equal closely the original mass of the quantity sieved.

HYDROMETER AND SIEVE ANALYSIS OF PORTION PASSING THE NO. 10 (2.00-mm) SIEVE

6. Determination of Composite Correction for Hydrometer Reading

6.1 Equations for percentages of soil remaining in suspension, as given in 13.3, are based on the use of distilled or demineralized water. A dispersing agent is used in the water, however, and the specific gravity of the resulting liquid is appreciably greater than that of distilled or demineralized water.

6.1.1 Both soil hydrometers are calibrated at 68 F (20 C), and variations in temperature from this standard temperature produce inaccuracies in the actual hydrometer readings. The amount of the inaccuracy increases as the variation from the standard temperature increases.

6.1.2 Hydrometers are graduated by the manufacturer to be read at the bottom of the meniscus formed by the liquid on the stem. Since it is not possible to secure readings of soil suspensions at the bottom of the meniscus, readings must be taken at the top and a correction applied.

6.1.3 The net amount of the corrections for the three items enumerated is designated as the composite correction, and may be determined experimentally.

6.2 For convenience, a graph or table of composite corrections for a series of 1-deg temperature differences for the range of expected test temperatures may be prepared and used as needed. Measurement of the composite corrections may be made at two temperatures spanning the range of expected test temperatures, and corrections for the intermediate temperatures calculated assuming a straight-line relationship between the two observed values.

6.3 Prepare 1000 ml of liquid composed of distilled or demineralized water and dispersing agent in the same proportion as will prevail in the sedimentation (hydrometer) test. Place the liquid in a sedimentation cylinder and the cylinder in the constant-temperature water bath, set for one of the two temperatures to be used. When the tempera-

ASTM **D 422**

ture of the liquid becomes constant, insert the hydrometer, and, after a short interval to permit the hydrometer to come to the temperature of the liquid, read the hydrometer at the top of the meniscus formed on the stem. For hydrometer 151H the composite correction is the difference between this reading and one; for hydrometer 152H it is the difference between the reading and zero. Bring the liquid and the hydrometer to the other temperature to be used, and secure the composite correction as before.

7. Hygroscopic Moisture

7.1 When the sample is weighed for the hydrometer test, weigh out an auxiliary portion of from 10 to 15 g in a small metal or glass container, dry the sample to a constant mass in an oven at 230 ± 9 F (110 ± 5 C), and weigh again. Record the masses.

8. Dispersion of Soil Sample

8.1 When the soil is mostly of the clay and silt sizes, weigh out a sample of air-dry soil of approximately 50 g. When the soil is mostly sand the sample should be approximately 100 g.

8.2 Place the sample in the 250-ml beaker and cover with 125 ml of sodium hexametaphosphate solution (40 g/litre). Stir until the soil is thoroughly wetted. Allow to soak for at least 16 h.

8.3 At the end of the soaking period, disperse the sample further, using either stirring apparatus A or B. If stirring apparatus A is used, transfer the soil - water slurry from the beaker into the special dispersion cup shown in Fig. 2, washing any residue from the beaker into the cup with distilled or demineralized water (Note 9). Add distilled or demineralized water, if necessary, so that the cup is more than half full. Stir for a period of 1 min.

NOTE 9—A large size syringe is a convenient device for handling the water in the washing operation. Other devices include the wash-water bottle and a hose with nozzle connected to a pressurized distilled water tank.

8.4 If stirring apparatus B (Fig. 3) is used, remove the cover cap and connect the cup to a compressed air supply by means of a rubber hose. An air gage must be on the line between the cup and the control valve. Open the control valve so that the gage indicates 1 psi (7

kPa) pressure (Note 10). Transfer the soil-water slurry from the beaker to the air-jet dispersion cup by washing with distilled or demineralized water. Add distilled or demineralized water, if necessary, so that the total volume in the cup is 250 ml, but no more.

NOTE 10—The initial air pressure of 1 psi is required to prevent the soil - water mixture from entering the air-jet chamber when the mixture is transferred to the dispersion cup.

8.5 Place the cover cap on the cup and open the air control valve until the gage pressure is 20 psi (140 kPa). Disperse the soil according to the following schedule:

Plasticity Index	Dispersion Period, min
Under 5	5
6 to 20	10
Over 20	15

Soils containing large percentages of mica need be dispersed for only 1 min. After the dispersion period, reduce the gage pressure to 1 psi preparatory to transfer of soil - water slurry to the sedimentation cylinder.

9. Hydrometer Test

9.1 Immediately after dispersion, transfer the soil - water slurry to the glass sedimentation cylinder, and add distilled or demineralized water until the total volume is 1000 ml.

9.2 Using the palm of the hand over the open end of the cylinder (or a rubber stopper in the open end), turn the cylinder upside down and back for a period of 1 min to complete the agitation of the slurry (Note 11). At the end of 1 min set the cylinder in a convenient location and take hydrometer readings at the following intervals of time (measured from the beginning of sedimentation), or as many as may be needed, depending on the sample or the specification for the material under test: 2, 5, 15, 30, 60, 250, and 1440 min. If the controlled water bath is used, the sedimentation cylinder should be placed in the bath between the 2- and 5-min readings.

NOTE 11—The number of turns during this minute should be approximately 60, counting the turn upside down and back as two turns. Any soil remaining in the bottom of the cylinder during the first few turns should be loosened by vigorous shaking of the cylinder while it is in the inverted position.

9.3 When it is desired to take a hydrometer reading, carefully insert the hydrometer about

D 422

20 to 25 s before the reading is due to approximately the depth it will have when the reading is taken. As soon as the reading is taken, carefully remove the hydrometer and place it with a spinning motion in a graduate of clean distilled or demineralized water.

NOTE 12—It is important to remove the hydrometer immediately after each reading. Readings shall be taken at the top of the meniscus formed by the suspension around the stem, since it is not possible to secure readings at the bottom of the meniscus.

9.4 After each reading, take the temperature of the suspension by inserting the thermometer into the suspension.

10. Sieve Analysis

10.1 After taking the final hydrometer reading, transfer the suspension to a No. 200 (75-μm) sieve and wash with tap water until the wash water is clear. Transfer the material on the No. 200 sieve to a suitable container, dry in an oven at 230 ± 9 F (110 ± 5 C) and make a sieve analysis of the portion retained, using as many sieves as desired, or required for the material, or upon the specification of the material under test.

CALCULATIONS AND REPORT

11. Sieve Analysis Values for the Portion Coarser than the No. 10 (2.00-mm) Sieve

11.1 Calculate the percentage passing the No. 10 sieve by dividing the mass passing the No. 10 sieve by the mass of soil originally split on the No. 10 sieve, and multiplying the result by 100. To obtain the mass passing the No. 10 sieve, subtract the mass retained on the No. 10 sieve from the original mass.

11.2 To secure the total mass of soil passing the No. 4 (4.75-mm) sieve, add to the mass of the material passing the No. 10 sieve the mass of the fraction passing the No. 4 sieve and retained on the No. 10 sieve. To secure the total mass of soil passing the ⅜-in. (9.5-mm) sieve, add to the total mass of soil passing the No. 4 sieve, the mass of the fraction passing the ⅜-in. sieve and retained on the No. 4 sieve. For the remaining sieves, continue the calculations in the same manner.

11.3 To determine the total percentage passing for each sieve, divide the total mass passing (see 11.2) by the total mass of sample and multiply the result by 100.

12. Hygroscopic Moisture Correction Factor

12.1 The hygroscopic moisture correction factor is the ratio between the mass of the oven-dried sample and the air-dry mass before drying. It is a number less than one, except when there is no hygroscopic moisture.

13. Percentages of Soil in Suspension

13.1 Calculate the oven-dry mass of soil used in the hydrometer analysis by multiplying the air-dry mass by the hygroscopic moisture correction factor.

13.2 Calculate the mass of a total sample represented by the mass of soil used in the hydrometer test, by dividing the oven-dry mass used by the percentage passing the No. 10 (2.00-mm) sieve, and multiplying the result by 100. This value is the weight W in the equation for percentage remaining in suspension.

13.3 The percentage of soil remaining in suspension at the level at which the hydrometer is measuring the density of the suspension may be calculated as follows (Note 13):
For hydrometer 151H:

$$P = [(100,000/W) \times G/(G - G_1)] (R - G_1)$$

NOTE 13—The bracketed portion of the equation for hydrometer 151H is constant for a series of readings and may be calculated first and then multiplied by the portion in the parenthesis.

For hydrometer 152H:

$$P = (Ra/W) \times 100$$

where:

a = correction faction to be applied to the reading of hydrometer 152H. (Values shown on the scale are computed using a specific gravity of 2.65. Correction factors are given in Table 1),

P = percentage of soil remaining in suspension at the level at which the hydrometer measures the density of the suspension,

R = hydrometer reading with composite correction applied (Section 6),

W = oven-dry mass of soil in a total test sample represented by mass of soil dispersed (see 13.2), g,

G = specific gravity of the soil particles, and

G_1 = specific gravity of the liquid in which soil particles are suspended. Use nu-

ASTM   **D 422**

merical value of one in both instances in the equation. In the first instance any possible variation produces no significant effect, and in the second instance, the composite correction for R is based on a value of one for G_1.

Diameter of Soil Particles

14.1 The diameter of a particle corresponding to the percentage indicated by a given hydrometer reading shall be calculated according to Stokes' law (Note 14), on the basis that a particle of this diameter was at the surface of the suspension at the beginning of sedimentation and had settled to the level at which the hydrometer is measuring the density of the suspension. According to Stokes' law:

$$D = \sqrt{[30n/980(G - G_1)] \times L/T}$$

where:

D = diameter of particle, mm,
n = coefficient of viscosity of the suspending medium (in this case water) in poises (varies with changes in temperature of the suspending medium),
L = distance from the surface of the suspension to the level at which the density of the suspension is being measured, cm. (For a given hydrometer and sedimentation cylinder, values vary according to the hydrometer readings. This distance is known as effective depth (Table 2),
T = interval of time from beginning of sedimentation to the taking of the reading, min,
G = specific gravity of soil particles, and
G_1 = specific gravity (relative density) of suspending medium (value may be used as 1.000 for all practical purposes).

NOTE 14—Since Stokes' law considers the terminal velocity of a single sphere falling in an infinity of liquid, the sizes calculated represent the diameter of spheres that would fall at the same rate as the soil particles.

14.2 For convenience in calculations the above equation may be written as follows:

$$D = K \sqrt{L/T}$$

where:

K = constant depending on the temperature of the suspension and the specific

gravity of the soil particles. Values of K for a range of temperatures and specific gravities are given in Table 3. The value of K does not change for a series of readings constituting a test, while values of L and T do vary.

14.3 Values of D may be computed with sufficient accuracy, using an ordinary 10-in. slide rule.

NOTE 15—The value of L is divided by T using the A- and B-scales, the square root being indicated on the D-scale. Without ascertaining the value of the square root it may be multiplied by K, using either the C- or CI-scale.

15. Sieve Analysis Values for Portion Finer than No. 10 (2.00-mm) Sieve

15.1 Calculation of percentages passing the various sieves used in sieving the portion of the sample from the hydrometer test involves several steps. The first step is to calculate the mass of the fraction that would have been retained on the No. 10 sieve had it not been removed. This mass is equal to the total percentage retained on the No. 10 sieve (100 minus total percentage passing) times the mass of the total sample represented by the mass of soil used (as calculated in 13.2), and the result divided by 100.

15.2 Calculate next the total mass passing the No. 200 sieve. Add together the fractional masses retained on all the sieves, including the No. 10 sieve, and subtract this sum from the mass of the total sample (as calculated in 13.2).

15.3 Calculate next the total masses passing each of the other sieves, in a manner similar to that given in 11.2.

15.4 Calculate last the total percentages passing by dividing the total mass passing (as calculated in 15.3) by the total mass of sample (as calculated in 13.2), and multiply the result by 100.

16. Graph

16.1 When the hydrometer analysis is performed, a graph of the test results shall be made, plotting the diameters of the particles on a logarithmic scale as the abscissa and the percentages smaller than the corresponding diameters to an arithmetic scale as the ordinate. When the hydrometer analysis is not

D 422

made on a portion of the soil, the preparation of the graph is optional, since values may be secured directly from tabulated data.

17. Report

17.1 The report shall include the following:

17.1.1 Maximum size of particles,

17.1.2 Percentage passing (or retained on) each sieve, which may be tabulated or presented by plotting on a graph (Note 16),

17.1.3 Description of sand and gravel particles:

17.1.3.1 Shape—rounded or angular,

17.1.3.2 Hardness—hard and durable, soft, or weathered and friable,

17.1.4 Specific gravity, if unusually high or low,

17.1.5 Any difficulty in dispersing the fraction passing the No. 10 (2.00-mm) sieve, indicating any change in type and amount of dispersing agent, and

17.1.6 The dispersion device used and the length of the dispersion period.

Note 16—This tabulation of graph represents the gradation of the sample tested. If particles larger than those contained in the sample were removed before testing, the report shall so state giving the amount and maximum size.

17.2 For materials tested for compliance with definite specifications, the fractions called for in such specifications shall be reported. The fractions smaller than the No. 10 sieve shall be read from the graph.

17.3 For materials for which compliance with definite specifications is not indicated and when the soil is composed almost entirely of particles passing the No. 4 (4.75-mm) sieve, the results read from the graph may be reported as follows:

(1) Gravel, passing 3-in. and retained on No. 4 sieve ..... percent

(2) Sand, passing No. 4 sieve and retained on No. 200 sieve ..... percent

　(a) Coarse sand, passing No. 4 sieve and retained on No. 10 sieve ..... percent

　(b) Medium sand, passing No. 10 sieve and retained on No. 40 sieve ..... percent

　(c) Fine sand, passing No. 40 sieve and retained on No. 200 sieve ..... percent

(3) Silt size, 0.074 to 0.005 mm ..... percent

(4) Clay size, smaller than 0.005 mm ..... percent

　Colloids, smaller than 0.001 mm ..... percent

17.4 For materials for which compliance with definite specifications is not indicated and when the soil contains material retained on the No. 4 sieve sufficient to require a sieve analysis on that portion, the results may be reported as follows (Note 17):

SIEVE ANALYSIS

Sieve Size	Percentage Passing
3-in.	....
2-in.	....
1½-in.	....
1-in.	....
¾-in.	....
⅜-in.	....
No. 4 (4.75-mm)	....
No. 10 (2.00-mm)	....
No. 40 (425-μm)	....
No. 200 (75-μm)	....

HYDROMETER ANALYSIS

0.074 mm	....
0.005 mm	....
0.001 mm	....

Note 17—No. 8 (2.36-mm) and No. 50 (300-μm) sieves may be substituted for No. 10 and No. 40 sieves.

ASTM   **D 422**

TABLE 1   Values of Correction Factor, *a*, for Different Specific Gravities of Soil Particles[a]

Specific Gravity	Correction Factor[a]
2.95	0.94
2.90	0.95
2.85	0.96
2.80	0.97
2.75	0.98
2.70	0.99
2.65	1.00
2.60	1.01
2.55	1.02
2.50	1.03
2.45	1.05

[a] For use in equation for percentage of soil remaining in suspension when using Hydrometer 152H.

TABLE 2   Values of Effective Depth Based on Hydrometer and Sedimentation Cylinder of Specified Sizes[a]

Hydrometer 151H		Hydrometer 152H			
Actual Hydrometer Reading	Effective Depth, L, cm	Actual Hydrometer Reading	Effective Depth, L, cm	Actual Hydrometer Reading	Effective Depth, L cm
1.000	16.3	0	16.3	31	11.2
1.001	16.0	1	16.1	32	11.1
1.002	15.8	2	16.0	33	10.9
1.003	15.5	3	15.8	34	10.7
1.004	15.2	4	15.6	35	10.6
1.005	15.0	5	15.5		
1.006	14.7	6	15.3	36	10.4
1.007	14.4	7	15.2	37	10.2
1.008	14.2	8	15.0	38	10.1
1.009	13.9	9	14.8	39	9.9
1.010	13.7	10	14.7	40	9.7
1.011	13.4	11	14.5	41	9.6
1.012	13.1	12	14.3	42	9.4
1.013	12.9	13	14.2	43	9.2
1.014	12.6	14	14.0	44	9.1
1.015	12.3	15	13.8	45	8.9
1.016	12.1	16	13.7	46	8.8
1.017	11.8	17	13.5	47	8.6
1.018	11.5	18	13.3	48	8.4
1.019	11.3	19	13.2	49	8.3
1.020	11.0	20	13.0	50	8.1
1.021	10.7	21	12.9	51	7.9
1.022	10.5	22	12.7	52	7.8
1.023	10.2	23	12.5	53	7.6
1.024	10.0	24	12.4	54	7.4
1.025	9.7	25	12.2	55	7.3
1.026	9.4	26	12.0	56	7.1
1.027	9.2	27	11.9	57	7.0
1.028	8.9	28	11.7	58	6.8
1.029	8.6	29	11.5	59	6.6
1.030	8.4	30	11.4	60	6.5

Table 2   *Continued*

Hydrometer 151H		Hydrometer 152H	
Actual Hydrometer Reading	Effective Depth, L, cm	Actual Hydrometer Reading	Effective Depth, L, cm
1.031	8.1		
1.032	7.8		
1.033	7.6		
1.034	7.3		
1.035	7.0		
1.036	6.8		
1.037	6.5		
1.038	6.2		

[a] Values of effective depth are calculated from the equation:

$$L = L_1 + \tfrac{1}{2}\,[L_2 - (V_B/A)]$$

where:
L = effective depth, cm,
L_1 = distance along the stem of the hydrometer from the top of the bulb to the mark for a hydrometer reading, cm,
L_2 = overall length of the hydrometer bulb, cm,
V_B = volume of hydrometer bulb, cm³, and
A = cross-sectional area of sedimentation cylinder, cm²
Values used in calculating the values in Table 2 are as follows:
For both hydrometers, 151H and 152H:
L_2 = 14.0 cm
V_B = 67.0 cm³
A = 27.8 cm²
For hydrometer 151H:
L_1 = 10.5 cm for a reading of 1.000
  = 2.3 cm for a reading of 1.031
For hydrometer 152H:
L_1 = 10.5 cm for a reading of 0 g/litre
  = 2.3 cm for a reading of 50 g/litre

ASTM   **D 422**

TABLE 3   Values of *K* for Use in Equation for Computing Diameter of Particle in Hydrometer Analysis

Temperature, deg C	Specific Gravity of Soil Particles								
	2.45	2.50	2.55	2.60	2.65	2.70	2.75	2.80	2.85
16	0.01510	0.01505	0.01481	0.01457	0.01435	0.01414	0.01394	0.01374	0.01356
17	0.01511	0.01486	0.01462	0.01439	0.01417	0.01396	0.01376	0.01356	0.01338
18	0.01492	0.01467	0.01443	0.01421	0.01399	0.01378	0.01359	0.01339	0.01321
19	0.01474	0.01449	0.01425	0.01403	0.01382	0.01361	0.01342	0.01323	0.01305
20	0.01456	0.01431	0.01408	0.01386	0.01365	0.01344	0.01325	0.01307	0.01289
21	0.01438	0.01414	0.01391	0.01369	0.01348	0.01328	0.01309	0.01291	0.01273
22	0.01421	0.01397	0.01374	0.01353	0.01332	0.01312	0.01294	0.01276	0.01258
23	0.01404	0.01381	0.01358	0.01337	0.01317	0.01297	0.01279	0.01261	0.01243
24	0.01388	0.01365	0.01342	0.01321	0.01301	0.01282	0.01264	0.01246	0.01229
25	0.01372	0.01349	0.01327	0.01306	0.01286	0.01267	0.01249	0.01232	0.01215
26	0.01357	0.01334	0.01312	0.01291	0.01272	0.01253	0.01235	0.01218	0.01201
27	0.01342	0.01319	0.01297	0.01277	0.01258	0.01239	0.01221	0.01204	0.01188
28	0.01327	0.01304	0.01283	0.01264	0.01244	0.01225	0.01208	0.01191	0.01175
29	0.01312	0.01290	0.01269	0.01249	0.01230	0.01212	0.01195	0.01178	0.01162
30	0.01298	0.01276	0.01256	0.01236	0.01217	0.01199	0.01182	0.01165	0.01149

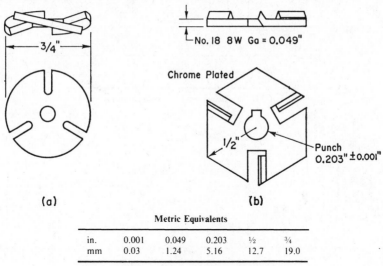

(a)   (b)

Metric Equivalents

in.	0.001	0.049	0.203	½	¾
mm	0.03	1.24	5.16	12.7	19.0

FIG. 1   Detail of Stirring Paddles.

ASTM   **D 422**

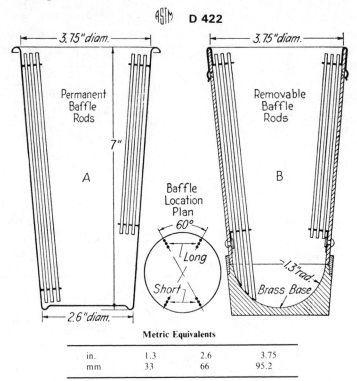

Metric Equivalents

in.	1.3	2.6	3.75
mm	33	66	95.2

FIG. 2   Dispersion Cups of Apparatus A.

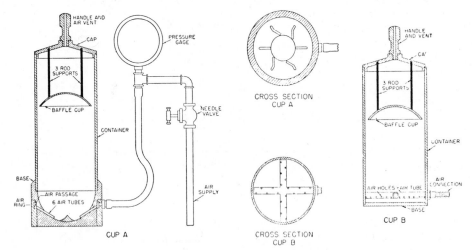

FIG. 3   Air-Jet Dispersion Cups of Apparatus B.

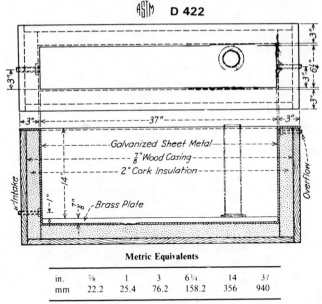

Metric Equivalents

in.	⅞	1	3	6¼	14	31
mm	22.2	25.4	76.2	158.2	356	940

FIG. 4   Insulated Water Bath.

The American Society for Testing and Materials takes no position respecting the validity of any patent rights asserted in connection with any item mentioned in this standard. Users of this standard are expressly advised that determination of the validity of any such patent rights, and the risk of infringement of such rights, is entirely their own responsibility.

This standard is subject to revision at any time by the responsible technical committee and must be reviewed every five years and if not revised, either reapproved or withdrawn. Your comments are invited either for revision of this standard or for additional standards and should be addressed to ASTM Headquarters. Your comments will receive careful consideration at a meeting of the responsible technical committee, which you may attend. If you feel that your comments have not received a fair hearing you should make your views known to the ASTM Committee on Standards, 1916 Race St., Philadelphia, Pa. 19103, which will schedule a further hearing regarding your comments. Failing satisfaction there, you may appeal to the ASTM Board of Directors.

ASTM D 423 – 66 (Reapproved 1972)

Standard Test Method for
LIQUID LIMIT OF SOILS[1]

This standard is issued under the fixed designation D 423; the number immediately following the designation indicates the year of original adoption or, in the case of revision, the year of last revision. A number in parentheses indicates the year of last reapproval.

1. Scope

1.1 This method covers the determination of the liquid limit of soils as defined in Section 2, using the mechanical device specified in Section 3, securing the results of at least three trials, and the plotting of a flow curve. Provision is also made for: (*1*) a one-point method requiring the calculation of the liquid limit value from data secured from a single trial; and (*2*) a procedure for check or referee tests.

1.2 The mechanical method shall be used unless specifications for the soil being tested permit the use of the one-point method. The procedure for check or referee tests shall be used in cases of disagreement or when required in the specifications for the soil being tested.

2. Definition

2.1 *liquid limit of soil* – the water content, expressed as a percentage of the weight of the oven-dried soil, at the boundary between the liquid and plastic states. The water content at this boundary is arbitrarily defined as the water content at which two halves of a soil cake will flow together for a distance of ½ in. (12.7 mm) along the bottom of the groove separating the two halves, when the cup is dropped 25 times for a distance of 1 cm (0.3937 in.) at the rate of 2 drops/s.

MECHANICAL METHOD

3. Apparatus

3.1 *Evaporating Dish*—A porcelain evaporating dish about 4½ in. (114.3 mm) in diameter.

3.2 *Spatula*—A spatula or pill knife having

a blade about 3 in. (76.2 mm) in length and about ¾ in. (19.0 mm) in width.

3.3 *Liquid Limit Device*—A mechanical device consisting of a brass cup and carriage, constructed according to the plan and dimensions shown in Fig. 1.[2]

3.4 *Grooving Tool*—A combined grooving tool and gage conforming to the dimensions shown in Fig. 1.

3.5 *Containers*—Suitable containers, such as matched watch glasses, that will prevent loss of moisture during weighing.

3.6 *Balance*—A balance sensitive to 0.01 g.

4. Sample

4.1 Take a sample weighing about 100 g from the thoroughly mixed portion of the material passing the No. 40 (425-μm) sieve, obtained in accordance with ASTM Method D 421, Dry Preparation of Soil Samples for Particle-Size Analysis and Determination of Soil Constants.[3]

5. Adjustment of Mechanical Device

5.1 Inspect the liquid-limit device to determine that the device is in good working order, that the pin connecting the cup is not worn sufficiently to permit side play, that the screws connecting the cup to the hanger arm are tight, and that a groove has not been worn in the cup through long usage. Inspect the

---

[1] This method is under the jurisdiction of ASTM Committee D-18 on Soil and Rock.
   Current edition approved Sept. 20, 1966. Originally issued 1935. Replaces D 423 – 61.
[2] Detailed working drawings for this liquid limit device are available at a nominal cost from the American Society for Testing and Materials, 1916 Race St., Philadelphia, Pa. 19103. Order Adjunct No. 12-404230-00.
[3] *Annual Book of ASTM Standards*, Part 19.

grooving tool to determine that the critical dimensions are as shown in Fig. 1.

5.2 By means of the gage on the handle of the grooving tool and the adjustment plate *H*, Fig. 1, adjust the height to which the cup *C* is lifted so that the point on the cup that comes in contact with the base is exactly 1 cm (0.3937 in.) above the base. Secure the adjustment plate *H* by tightening the screws, *I*. With the gage still in place, check the adjustment by revolving the crank rapidly several times. If the adjustment is correct, a slight ringing sound will be heard when the cam strikes the cam follower. If the cup is raised off the gage or no sound is heard, make further adjustments.

6. Procedure

6.1 Place the soil sample in the evaporating dish and thoroughly mix with 15 to 20 ml of distilled water by alternately and repeatedly stirring, kneading, and chopping with a spatula. Make further additions of water in increments of 1 to 3 ml. Thoroughly mix each increment of water with the soil, as previously described, before adding another increment of water.

6.2 When sufficient water has been thoroughly mixed with the soil to produce a consistency that will require 30 to 35 drops of the cup to cause closure, place a portion of the mixture in the cup above the spot where the cup rests on the base, and squeeze it down and spread it into the position shown in Fig. 2 with as few strokes of the spatula as possible, care being taken to prevent the entrapment of air bubbles within the mass. With the spatula, level the soil and at the same time trim it to a depth of 1 cm at the point of maximum thickness. Return the excess soil to the evaporating dish. Divide the soil in the cup by firm strokes of the grooving tool along the diameter through the centerline of the cam follower so that a clean, sharp groove of the proper dimensions will be formed. To avoid tearing of the sides of the groove or slipping of the soil cake on the cup, up to six strokes, from front to back or from back to front counting as one stroke, shall be permitted. Each stroke should penetrate a little deeper until the last stroke from back to front scrapes the bottom of the cup clean. Make the groove with as few strokes as possible.

6.3 Lift and drop the cup by turning the crank, *F*, at the rate of 2 rps, until the two halves of the soil cake come in contact at the bottom of the groove along a distance of about ½ in. (12.7 mm). Record the number of drops required to close the groove along a distance of ½ in.

6.4 Remove a slice of soil approximately the width of the spatula, extending from edge to edge of the soil cake at right angles to the groove and including that portion of the groove in which the soil flowed together, and place in a suitable tared container. Weigh and record the mass. Oven-dry the soil in the container to constant mass at 230 ± 9 F (110 ± 5 C) and reweigh as soon as it has cooled but before hygroscopic moisture can be absorbed. Record this mass. Record the loss in mass due to drying as the mass of water.

6.5 Transfer the soil remaining in the cup to the evaporating dish. Wash and dry the cup and grooving tool, and reattach the cup to the carriage in preparation for the next trial.

6.6 Repeat the foregoing operations for at least two additional trials, with the soil collected in the evaporating dish, to which sufficient water has been added to bring the soil to a more fluid condition. The object of this procedure is to obtain samples of such consistency that the number of drops required to close the groove will be above and below 25. The number of drops should be less than 35 and exceed 15. The test shall always proceed from the dryer to the wetter condition of the soil.

7. Calculation

7.1 Calculate the water content w_N of the soil, expressed as a percentage of the weight of the oven-dried soil, as follows:

$$w_N = \text{(mass of water/mass of oven-dried soil)} \times 100$$

8. Preparation of the Flow Curve

8.1 Plot a "flow curve" representing the relationship between water content and corresponding numbers of drops of the cup on a semilogarithmic graph with the water content as abscissae on the arithmetical scale, and the number of drops as ordinates on the logarithmic scale. The flow curve is a straight line drawn as nearly as possible through the three or more plotted points.

ASTM   **D 423**

9. Liquid Limit

9.1 Take the water content corresponding to the intersection of the flow curve with the 25-drop ordinate as the liquid limit of the soil. Report this value to the nearest whole number.

ONE-POINT METHOD

10. Apparatus

10.1 The requirements for apparatus are the same as specified in Section 3.

11. Sample

11.1 The requirements for the sample are the same as specified in Section 4.

12. Adjustment of Mechanical Device

12.1 The requirements for the mechanical device are the same as specified in Section 5.

13. Procedure

13.1 Proceed in accordance with 6.1 through 6.5, except that a moisture content sample shall be taken only for the accepted trial. The accepted trial shall require between 20 and 30 drops of the cup to close the groove and at least two consistent consecutive closures shall be observed before taking the moisture-content sample for calculation of the liquid limit. The test shall always proceed from the dryer to the wetter condition of the soil.

14. Calculations

14.1 Calculate the water content w_N, for the accepted trial, expressed as a percentage of the oven-dried weight, as follows:

$w_N =$
   (mass of water/mass of oven-dried soil) $\times$ 100

14.2 Determine the liquid limit, LL, using the following formula:

$$LL = w_N(N/25)^{0.12}$$

where:
N = number of drops of the cup required to close the groove at the water content, w_N.

14.3 Report the liquid limit value to the nearest whole number. Values of $(N/25)^{0.12}$ are given in Table 1.

CHECK OR REFEREE TESTS

15. Factors Influencing Results

15.1 The results of liquid limit tests are influenced by:

15.1.1 The time required to make the test, and

15.1.2 The moisture content at which the test is begun.

16. Procedure

16.1 In making the liquid limit test for check or referee purposes, use the following time schedule:

16.1.1 Mixing of soil with water: 5 to 20 min, the longer period being used for the more plastic soils.

16.1.2 Seasoning in the humidifier: 24 h.

16.1.3 Remixing before placing in the brass cup: Add 1 ml of water and mix for 1 min. Add the water in 1-ml increments and remix until the soil appears to have such a consistency that the groove will close in 35 to 25 blows.

16.1.4 Placing in the brass cup, testing and weighing: 3 min.

16.1.5 Washing cup and grooving tool, adding water, and remixing: 3 min.

16.1.6 Repeat operations 4 and 5 for at least three satisfactory trials.

16.2 Do not record any trial requiring more than 35 or less than 15 drops of the cup. In no case shall dried soil be added to the seasoned soil being tested.

17. Report

17.1 The report shall include the following:

17.1.1 Liquid limit value, and

17.1.2 Procedure followed when either the one-point method or the procedure for check or referee tests is used.

ASTM   **D 423**

TABLE 1   Values of $(N/25)^{0.12}$

N	$(N/25)^{0.12}$	N	$(N/25)^{0.12}$	N	$(N/25)^{0.12}$
20	0.974	24	0.995	28	1.014
21	0.979	25	1.000	29	1.018
22	0.985	26	1.005	30	1.022
23	0.990	27	1.009		

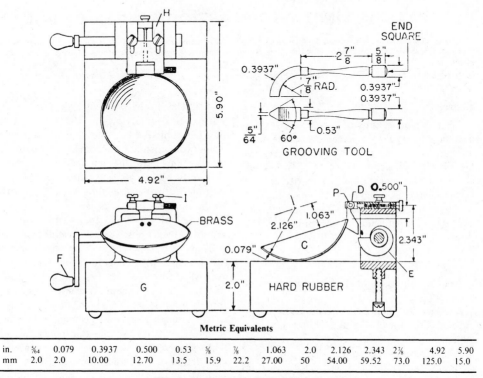

Metric Equivalents

in.	5/64	0.079	0.3937	0.500	0.53	5/8	7/8	1.063	2.0	2.126	2.343	2 7/8	4.92	5.90
mm	2.0	2.0	10.00	12.70	13.5	15.9	22.2	27.00	50	54.00	59.52	73.0	125.0	15.0

FIG. 1   Mechanical Liquid Limit Device.

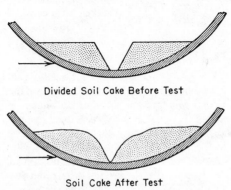

Divided Soil Cake Before Test

Soil Cake After Test

FIG. 2   Diagram Illustrating Liquid Limit Test.

ASTM D 424 – 59 (Reapproved 1971)

Standard Test Method for

PLASTIC LIMIT AND PLASTICITY INDEX OF SOILS[1]

This standard is issued under the fixed designation D 424; the number immediately following the designation indicates the year of original adoption or, in the case of revision, the year of last revision. A number in parentheses indicates the year of last reapproval.

This method has been approved for use by agencies of the Department of Defense and for listing in the DoD Index of Specifications and Standards.

1. Scope

1.1 The plastic limit of a soil is the water content, expressed as a percentage of the mass of oven-dried soil, at the boundary between the plastic and semisolid states. The water content at this boundary is arbitrarily defined as the lowest water content at which the soil can be rolled into threads ⅛ in. (3.2 mm) in diameter without the threads breaking into pieces.

2. Apparatus

2.1 The apparatus shall consist of the following:

2.1.1 *Evaporating Dish*—A porcelain evaporating dish about 4½ in. (114.3 mm) in diameter.

2.1.2 *Spatula*—A spatula or pill knife having a blade about 3 in. (76.2 mm) in length and about ¾ in. (19.0 mm) in width.

2.1.3 *Surface for Rolling*—A ground-glass plate or piece of glazed or unglazed paper on which to roll the sample.

NOTE 1—"Unglazed" refers to paper similar to that used for mimeographing. Paper toweling is not satisfactory.

2.1.4 *Containers*—Suitable containers, such as matched watch glasses, that will prevent loss of moisture during weighing.

2.1.5 *Balance*—A balance sensitive to 0.01 g.

3. Sample

3.1 If the plastic limit only is required, take a quantity of soil weighing about 15 g from the thoroughly mixed portion of the material passing the No. 40 (425-μm) sieve, obtained in accordance with ASTM Method

D 421, Dry Preparation of Soil Samples for Particle-Size Analysis and Determination of Soil Constants.[2] Place the air-dried soil in an evaporating dish and thoroughly mix with distilled water until the mass becomes plastic enough to be easily shaped into a ball. Take a portion of this ball weighing about 8 g for the test sample.

3.2 If both the liquid and plastic limits are required, take a test sample weighing about 8 g from the thoroughly wet and mixed portion of the soil prepared in accordance with ASTM Method D 423, Test for Liquid Limit of Soils.[2] Take the sample at any stage of the mixing process at which the mass becomes plastic enough to be easily shaped into a ball without sticking to the fingers excessively when squeezed. If the sample is taken before completion of the liquid limit test, set it aside and allow to season in air until the liquid limit test has been completed. If the sample is taken after completion of the liquid limit test, and is still too dry to permit rolling to a ⅛-in. (3.2-mm) thread, add more water.

4. Procedure

4.1 Squeeze and form the 8-g test sample taken in accordance with 3.1 or 3.2 into an ellipsoidal-shape mass. Roll this mass between the fingers and the ground-glass plate or a piece of paper lying on a smooth horizontal surface with just sufficient pressure to roll the mass into a thread of uniform diameter throughout its length. The rate of rolling shall be between 80 and 90 strokes/min,

---

[1] This method is under the jurisdiction of ASTM Committee D-18 on Soil and Rock.

Current edition approved Sept. 19, 1959. Originally issued 1935. Replaces D 424 – 54.

[2] *Annual Book of ASTM Standards*, Part 19.

ASTM   **D 424**

counting a stroke as one complete motion of the hand forward and back to the starting position again.

4.2 When the diameter of the thread becomes $1/8$ in. (3.2 mm), break the thread into six or eight pieces. Squeeze the pieces together between the thumbs and fingers of both hands into a uniform mass roughly ellipsoidal in shape, and reroll. Continue this alternate rolling to a thread $1/8$ in. in diameter, gathering together, kneading and rerolling, until the thread crumbles under the pressure required for rolling and the soil can no longer be rolled into a thread. The crumbling may occur when the thread has a diameter greater than $1/8$ in. This shall be considered a satisfactory end point, provided the soil has been previously rolled into a thread $1/8$ in. in diameter. The crumbling will manifest itself differently with the various types of soil. Some soils fall apart in numerous small aggregations of particles; others may form an outside tubular layer that starts splitting at both ends. The splitting progresses toward the middle, and finally, the thread falls apart in many small platy particles. Heavy clay soils require much pressure to deform the thread, particularly as they approach the plastic limit, and finally, the thread breaks into a series of barrel-shaped segments each about $1/4$ to $3/8$ in. (6.4 to 9.5 mm) in length. At no time shall the operator attempt to produce failure at exactly $1/8$-in. diameter by allowing the thread to reach $1/8$ in., then reducing the rate of rolling or the hand pressure, or both, and continuing the rolling without further deformation until the thread falls apart. It is permissible, however, to reduce the total amount of deforma-tion for feebly plastic soils by making the initial diameter of the ellipsoidal-shaped mass nearer to the required $1/8$ in. final diameter.

4.3 Gather the portions of the crumbled soil together and place in a suitable tared container. Weigh the container and soil and record the mass. Oven-dry the soil in the container to constant mass at 230 ± 9 F (110 ± 5 C) and weigh. Record this mass. Record the loss in mass as the mass of water.

5. Calculations

5.1 Calculate the plastic limit, expressed as the water content in percentage of the mass of the oven-dry soil, as follows:

Plastic limit
= (mass of water/mass of oven-dry soil) × 100

Report the plastic limit to the nearest whole number.

5.2 Calculate the plasticity index of a soil as the difference between its liquid limit and its plastic limit, as follows:

Plasticity index = liquid limit − plastic limit

5.3 Report the difference calculated as indicated in 5.2 as the plasticity index, except under the following conditions:

5.3.1 When the liquid limit or plastic limit cannot be determined, report the plasticity index as NP (nonplastic).

5.3.2 When the soil is extremely sandy, the plastic limit test shall be made before the liquid limit. If the plastic limit cannot be determined, report both the liquid limit and plastic limit as NP.

5.3.3 When the plastic limit is equal to, or greater than, the liquid limit, report the plasticity index as NP.

ASTM D 698 – 78

Standard Test Methods for

MOISTURE-DENSITY RELATIONS OF SOILS AND SOIL-AGGREGATE MIXTURES USING 5.5-lb (2.49-kg) RAMMER AND 12-in. (305-mm) DROP[1]

This standard is issued under the fixed designation D 698; the number immediately following the designation indicates the year of original adoption or, in the case of revision, the year of last revision. A number in parentheses indicates the year of last reapproval.

1. Scope

1.1 These laboratory compaction methods cover the determination of the relationship between the moisture content and density of soils and soil-aggregate mixtures (Note 1) when compacted in a mold of a given size with a 5.5-lb (2.49-kg) rammer dropped from a height of 12 in. (305 mm) (Note 2). Four alternative procedures are provided as follows:

1.1.1 *Method A*—A 4-in. (101.6-mm) mold; material passing a No. 4 (4.75-mm) sieve;

1.1.2 *Method B*—A 6-in. (152.4-mm) mold; material passing a No. 4 (4.75-mm) sieve;

1.1.3 *Method C*—A 6-in. (152.4-mm) mold; material passing a ¾-in. (19.0-mm) sieve; and

1.1.4 *Method D*—A 6-in. (152.4-mm) mold; material passing a ¾-in. (19.0-mm) sieve, corrected by replacement for material retained on a ¾-in. sieve.

NOTE 1—Soils and soil-aggregate mixtures should be regarded as natural occurring fine- or coarse-grained soils or composites or mixtures of natural soils, or mixtures of natural and processed soils or aggregates such as silt, gravel, or crushed rock.

NOTE 2—These laboratory compaction test methods when used on soils and soil-aggregates which are not free-draining will, in most cases, establish a well-defined optimum moisture content and maximum density (see Section 7). However, for free-draining soils and soil-aggregate mixtures, these methods will not, in many cases, produce a well-defined moisture-density relationship and the maximum density obtained will generally be less than that obtained by vibratory methods.

1.2 The method to be used should be indicated in the specifications for the material being tested. If no method is specified, the provisions of Section 5 shall govern.

2. Applicable Documents

2.1 *ASTM Standards*:

C 127 Test for Specific Gravity and Absorption of Coarse Aggregate[2]

D 854 Test for Specific Gravity of Soils[3]

D 2168 Calibration of Laboratory Mechanical-Rammer Soil Compactors[3]

D 2216 Test for Laboratory Determination of Water (Moisture) Content of Soil, Rock, and Soil-Aggregate Mixtures[3]

D 2487 Classification of Soils for Engineering Purposes[3]

D 2488 Recommended Practice for Description of Soils (Visual-Manual Procedure)[3]

E 11 Specification for Wire-Cloth Sieves for Testing Purposes[4]

3. Apparatus

3.1 *Molds*—The molds shall be cylindrical in shape, made of rigid metal and be within the capacity and dimensions indicated in 3.1.1 or 3.1.2. The molds may be the "split" type, consisting either of two half-round sections, or a section of pipe split along one element,

---

[1] These methods are under the jurisdiction of ASTM Committee D-18 on Soil and Rock.

Current edition approved April 27, 1978. Published July 1978. Originally published as D 698 – 42 T. Last previous edition D 698 – 70.

[2] *Annual Book of ASTM Standards*, Parts 14 and 15.

[3] *Annual Book of ASTM Standards*, Part 19.

[4] *Annual Book of ASTM Standards*, Parts 14, 15, and 41.

ASTM **D 698**

which can be securely locked together to form a cylinder meeting the requirements of this section. The molds may also be the "taper" type, providing the internal diameter taper is uniform and is not more than 0.200 in./linear ft (16.7 mm/linear m) of mold height. Each mold shall have a base plate assembly and an extension collar assembly, both made of rigid metal and constructed so they can be securely attached to or detached from the mold. The extension collar assembly shall have a height extending above the top of the mold of at least 2 in. (50.8 mm), which may include an upper section that flares out to form a funnel provided there is at least a $^3/_4$-in. (19-mm) straight cylindrical section beneath it.

3.1.1 *Mold*, 4.0 in. (101.6 mm) in diameter, having a capacity of $^1/_{30} \pm 0.0004$ ft^3 (944 $\pm$ 11 cm^3) and conforming to Fig. 1.

3.1.2 *Mold*, 6.0 in. (152.4 mm) in diameter, having a capacity of $^1/_{13.333} \pm 0.0009$ ft^3 (2124 $\pm$ 25 cm^3) and conforming to Fig. 2.

3.1.3 The average internal diameter, height, and volume of each mold shall be determined before initial use and at intervals not exceeding 1000 times the mold is filled. The mold volume shall be calculated from the average of at least six internal diameter and three height measurements made to the nearest 0.001 in. (0.02 mm), or from the amount of water required to completely fill the mold, corrected for temperature variance in accordance with Table 1. If the average internal diameter and volume are not within the tolerances shown in Figs. 1 or 2, the mold shall not be used. The determined volume shall be used in computing the required densities.

3.2 *Rammer* —The rammer may be either manually operated (see 3.2.1) or mechanically operated (see 3.2.2). The rammer shall fall freely through a distance of 12.0 $\pm$ $^1/_{16}$ in. (304.8 $\pm$ 1.6 mm) from the surface of the specimen. The manufactured weight of the rammer shall be 5.5 $\pm$ 0.02 lb (2.49 $\pm$ 0.01 kg). The specimen contact face shall be flat.

3.2.1 *Manual Rammer*—The specimen contact face shall be circular with a diameter of 2.000 $\pm$ 0.005 in. (50.80 $\pm$ 0.13 mm). The rammer shall be equipped with a guide-sleeve which shall provide sufficient clearance so that the free fall of the rammer shaft and head will not be restricted. The guidesleeve shall have four vent holes at each end (eight

holes total) located with centers $^3/_4$ $\pm$ $^1/_{16}$ in. (19.0 $\pm$ 1.6 mm) from each end and spaced 90 deg apart. The minimum diameter of the vent holes shall be $^3/_8$ in. (9.5 mm).

3.2.2 *Mechanical Rammer* —The rammer shall operate mechanically in such a manner as to provide uniform and complete coverage of the specimen surface. There shall be 0.10 $\pm$ 0.03 in. (2.5 $\pm$ 0.8 mm) clearance between the rammer and the inside surface of the mold at its smallest diameter. When used with the 4.0-in. (101.6-mm) mold, the specimen contact face shall be circular with a diameter of 2.000 $\pm$ 0.005 in. (50.80 $\pm$ 0.13 mm). When used with the 6.0-in. (152.4-mm) mold, the specimen contact face shall have the shape of a section of a circle of a radius equal to 2.90 $\pm$ 0.02 in. (73.7 $\pm$ 0.5 mm). The sector face rammer shall operate in such a manner that the vertex of the sector is positioned at the center of the specimen. The mechanical rammer shall be calibrated and adjusted, as necessary, in accordance with 3.2.3.

3.2.3 *Calibration and Adjustment* —The mechanical rammer shall be calibrated, and adjusted as necessary, before initial use; near the end of each period during which the mold was filled 1000 times; before reuse after anything, including repairs, which may affect the test results significantly; and whenever the test results are questionable. Each calibration and adjustment shall be in accordance with Method D 2168.

3.3 *Sample Extruder* (optional) —A jack, frame or other device adapted for the purpose of extruding compacted specimens from the mold.

3.4 *Balances* —A balance or scale of at least 20-kg capacity sensitive to $\pm$ 1 g and a balance of at least 1000-g capacity sensitive to $\pm$0.01 g.

3.5 *Drying Oven*, thermostatically controlled, preferably of the forced-draft type, capable of maintaining a temperature of 230 $\pm$ 9°F (110 $\pm$ 5°C) for determining the moisture content of the compacted specimen.

3.6 *Straightedge*—A stiff metal straightedge of any convenient length but not less than 10 in. (254 mm). The scraping edge shall have a straightness tolerance of $\pm$0.005 in. ($\pm$0.13 mm) and shall be beveled if it is thicker than $^1/_8$ in. (3 mm).

3.7 *Sieves*, 3-in. (75-mm), $^3/_4$-in. (19.0-

mm) and No. 4 (4.75-mm), conforming to the requirements of Specification E 11.

3.8 *Mixing Tools* — Miscellaneous tools such as mixing pan, spoon, trowel, spatula, etc., or a suitable mechanical device for thoroughly mixing the sample of soil with increments of water.

4. Procedure

4.1 *Specimen Preparation* — Select a representative portion of quantity adequate to provide, after sieving, an amount of material weighing as follows: Method A — 25 lb (11 kg); Methods B, C, and D — 50 lb (23 kg). Prepare specimens in accordance with either 4.1.1 through 4.1.3 or 4.1.4.

4.1.1 *Dry Preparation Procedure* — If the sample is too damp to be friable, reduce the moisture content by drying until the material is friable; see 4.1.2. Drying may be in air or by the use of a drying apparatus such that the temperature of the sample does not exceed 140°F (60°C). After drying (if required), thoroughly break up the aggregations in such a manner as to avoid reducing the natural size of the particles. Pass the material through the specified sieve as follows: Methods A and B — No. 4 (4.75-mm); Methods C and D — ³/₄-in. (19.0-mm). Correct for oversize material in accordance with Section 5, if Method D is specified.

4.1.2 Whenever practicable, soils classified as ML, CL, OL, GC, SC, MH, CH, OH and PT by Classification D 2487 shall be prepared in accordance with 4.1.4.

4.1.3 Prepare a series of at least four specimens by adding increasing amounts of water to each sample so that the moisture contents vary by approximately 1¹/₂ %. The moisture contents selected shall bracket the optimum moisture content, thus providing specimens which, when compacted, will increase in mass to the maximum density and then decrease in density (see 7.2 and 7.3). Thoroughly mix each specimen to ensure even distribution of moisture throughout and then place in a separate covered container and allow to stand prior to compaction in accordance with Table 2. For the purpose of selecting a standing time, it is not required to perform the actual classification procedures described in Classification D 2487 (except in the case of referee

testing), if previous data exist which provide a basis for classifying the sample.

4.1.4 *Moist Preparation Method* — The following alternate procedure is recommended for soils classified as ML, CL, OL, GC, SC, MH, CH, OH, and PT by Classification D 2487. Without previously drying the sample, pass it through the ³/₄-in. (19.0-mm) and No. 4 (4.75-mm) sieves. Correct for oversize material in accordance with Section 5, if Method D is specified. Prepare a series of at least four specimens having moisture contents that vary by approximately 1¹/₂ %. The moisture contents selected shall bracket the optimum moisture content, thus providing specimens which, when compacted, will increase in mass to the maximum density and then decrease in density (see 7.2 and 7.3). To obtain the appropriate moisture content of each specimen, the addition of a predetermined amount of water (see 4.1.3) or the removal of a predetermined amount of moisture by drying may be necessary. Drying may be in air or by the use of a drying apparatus such that the temperature of the specimen does not exceed 140°F (60°C). The prepared specimens shall then be thoroughly mixed and stand, as specified in 4.1.3 and Table 2, prior to compaction.

NOTE 3 — With practice, it is usually possible to visually judge the point of optimum moisture closely enough so that the prepared specimens will bracket the point of optimum moisture content.

4.2 *Specimen Compaction* — Select the proper compaction mold, in accordance with the method being used, and attach the mold extension collar. Compact each specimen in three layers of approximately equal height. Each layer shall receive 25 blows in the case of the 4-in. (101.6-mm) mold; each layer shall receive 56 blows in the case of the 6-in. (152.4-mm) mold. The total amount of material used shall be such that the third compacted layer is slightly above the top of the mold, but not exceeding ¹/₄ in. (6 mm). During compaction the mold shall rest on a uniform rigid foundation, such as provided by a cylinder or cube of concrete weighing not less than 200 lb (91 kg).

4.2.1 In operating the manual rammer, care shall be taken to avoid rebound of the rammer from the top end of the guidesleeve.

The guidesleeve shall be held steady and within 5 deg of the vertical. Apply the blows at a uniform rate not exceeding approximately 1.4 s per blow and in such a manner as to provide complete coverage of the specimen surface.

4.2.2 Following compaction, remove the extension collar; carefully trim the compacted specimen even with the top of the mold by means of the straightedge and determine the mass of the specimen. Divide the mass of the compacted specimen and mold, minus the mass of the mold, by the volume of the mold (see 3.1.3). Record the result as the wet density, γ_m, in pounds per cubic foot (or kilograms per cubic metre) of the compacted specimen.

4.2.3 Remove the material from the mold. Determine moisture content in accordance with Method D 2216, using either the whole specimen or alternatively a representative specimen of the whole specimen. The whole specimen must be used when the permeability of the compacted specimen is high enough so that the moisture content is not distributed uniformly throughout. If the whole specimen is used, break it up to facilitate drying. Obtain the representative specimen by slicing the compacted specimen axially through the center and removing 100 to 500 g of material from one of the cut faces.

4.2.4 Repeat 4.2 through 4.2.3 for each specimen prepared.

5. Oversize Corrections

5.1 If 30 % or more of the sample is retained on a $^3/_4$-in. (19.0-mm) sieve, then none of the methods described under these methods shall be used for the determination of either maximum density or optimum moisture content.

5.2 *Methods A and B*—The material retained on the No. 4 (4.75-mm) sieve is discarded and no oversize correction is made. However, it is recommended that if the amount of material retained is 7 % or greater, Method C be used instead.

5.3 *Method C*—The material retained on the $^3/_4$-in. (19.0-mm) sieve is discarded and no oversize correction is made. However, if the amount of material retained is 10 % or greater, it is recommended that Method D be used instead.

5.4 *Method D:*

5.4.1 This method shall not be used unless the amount of material retained on the $^3/_4$-in. (19.0-mm) sieve is 10 % or greater. When the amount of material retained on the $^3/_4$-in. sieve is less than 10 %, use Method C.

5.4.2 Pass the material retained on the $^3/_4$-in. (19.0-mm) sieve through a 3-in. or 75-mm sieve. Discard the material retained on the 3-in. sieve. The material passing the 3-in. sieve and retained on the $^3/_4$-in. sieve shall be replaced with an equal amount of material passing a $^3/_4$-in. sieve and retained on a No. 4 (4.75-mm) sieve. The material for replacement shall be taken from an unused portion of the sample.

6. Calculations

6.1 Calculate the moisture content and the dry density of each compacted specimen as follows:

$$w = [(A - B)/(B - C)] \times 100$$

and

$$\gamma_d = [\gamma_m/(w + 100)] \times 100$$

where:

w = moisture content in percent of the compacted specimens,

A = mass of container and moist specimen,

B = mass of container and oven-dried specimen,

C = mass of container,

γ_d = dry density, in pounds per cubic foot (or kilograms per cubic metre) of the compacted specimen, and

γ_m = wet density, in pounds per cubic foot (or kilograms per cubic metre) of the compacted specimen.

7. Moisture-Density Relationship

7.1 From the data obtained in 6.1, plot the dry density values as ordinates with corresponding moisture contents as abscissas. Draw a smooth curve connecting the plotted points. Also draw a curve termed the "curve of complete saturation" or "zero air voids curve" on this plot. This curve represents the relationship between dry density and corresponding moisture contents when the voids are completely filled with water. Values of dry density and corresponding moisture contents for plotting the curve of complete saturation can be com-

D 698

puted using the following equation:

$$w_{sat} = [(62.4/\gamma_d) - (1/G_s)] \times 100$$

where:

w_{sat} = moisture content in percent for complete saturation,

γ_d = dry density in pounds per cubic foot (or kilograms per cubic metre),

G_s = specific gravity of the material being tested (see Note 4), and

62.4 = density of water in pounds per cubic foot (or kilograms per cubic metre).

NOTE 4—The specific gravity of the material can either be assumed or based on the weighted average values of: (a) the specific gravity of the material passing the No. 4 (4.75-mm) sieve in accordance with Method D 854; and (b) the apparent specific gravity of the material retained on the No. 4 sieve in accordance with Method C 127.

7.2 *Optimum Moisture Content*, w_o—The moisture content corresponding to the peak of the curve drawn as directed in 7.1 shall be termed the "optimum moisture content."

7.3 *Maximum Density*, γ_{max}—The dry density in pounds per cubic foot (or kilograms per cubic metre) of the sample at "optimum moisture content" shall be termed "maximum density."

8. Report

8.1 The report shall include the following.

8.1.1 Method used (Method A, B, C, or D).

8.1.2 Optimum moisture content.

8.1.3 Maximum density.

8.1.4 Description of rammer (whether manual or mechanical).

8.1.5 Description of appearance of material used in test, based on Recommended Practice D 2488 (Classification D 2487 may be used as an alternative).

8.1.6 Origin of material used in test.

8.1.7 Preparation procedure used (moist or dry).

9. Precision

9.1 Criteria for judging the acceptability of the maximum density and optimum moisture content test results are given in Table 3. The standard deviation, s, is calculated from the equation:

$$s^2 = \frac{1}{n-1} \sum_1^n (x - \bar{x})^2$$

where:

n = number of determinations,

x = individual value of each determination, and

$\bar{x}$ = numerical average of the determinations. tions.

9.2 Criteria for assigning standard deviation values for single-operator precision are not available at the present time.

TABLE 1   Volume of Water per Gram based on Temperature[A]

Temperature, °C (°F)	Volume of Water, ml/g
12 (53.6)	1.00048
14 (57.2)	1.00073
16 (60.8)	1.00103
18 (64.4)	1.00138
20 (68.0)	1.00177
22 (71.6)	1.00221
24 (75.2)	1.00268
26 (78.8)	1.00320
28 (82.4)	1.00375
30 (86.0)	1.00435
32 (89.6)	1.00497

[A] Values other than shown may be obtained by referring to the *Handbook of Chemistry and Physics*, Chemical Rubber Publishing Co., Cleveland, Ohio.

TABLE 2   Dry Preparation Method—Standing Times

Classification D 2487	Minimum Standing Time, h
GW, GP, SW, SP	no requirement
GM, SM	3
ML, CL, OL, GC, SC	18
MH, CH, OH, PT	36

D 698

TABLE 3   Precision

	Standard Deviation, s	Acceptable Range of Two Results, Expressed as Percent of Mean Value[A]
Single-operator precision:		
Maximum density	...	1.9
Optimum moisture content	...	9.5
Multilaboratory precision:		
Maximum density	±1.66	4.0
Optimum moisture content	±0.86	15.0

[A] This column indicates a limiting range of values which should not be exceeded by the difference between any two results, expressed as a percentage of the average value. In cooperative test programs it has been determined that 95 % of the tests do not exceed the limiting acceptable ranges shown below. All values shown in this table are based on average test results from a variety of different soils and are subject to future revision.

TABLE 4   Metric Equivalents for Figs. 1 and 2

in.	mm
0.016	0.41
0.026	0.66
1/32	0.80
1/16	1.6
1/8	3.2
1/4	6.4
11/32	8.7
3/8	9.5
1/2	12.7
5/8	15.9
2	50.8
2 1/2	63.5
4	101.6
4 1/4	108.0
4 1/2	114.3
4.584	116.43
6	152.4
6 1/2	165.1
8	203.2

ft³	cm³
1/30	944
0.004	11
1/13.333	2124
0.0009	25

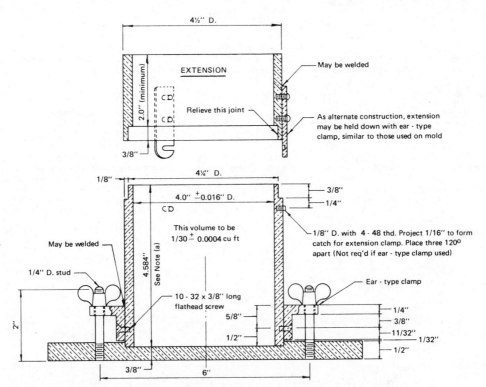

NOTE 1—The tolerance on the height is governed by the allowable volume and diameter tolerances.
NOTE 2—The methods shown for attaching the extension collar to the mold and the mold to the base plate are recommended. However, other methods are acceptable, providing the attachments are equally as rigid as those shown.
FIG. 1   Cylindrical Mold, 4.0-in. for Soil Tests (see Table 4 for metric equivalents).

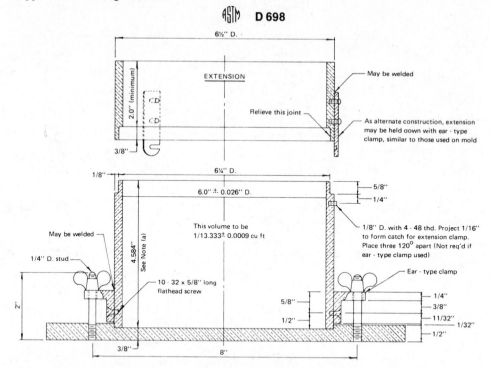

@STM **D 698**

NOTE 1 —The tolerance on the height is governed by the allowable volume and diameter tolerances.
NOTE 2 —The methods shown for attaching the extension collar to the mold and the mold to the base plate are recommended. However, other methods are acceptable, providing the attachments are equally as rigid as those shown.

FIG. 2   Cylindrical Mold, 6.0-in. for Soil Tests (see Table 4 for metric equivalents).

ASTM D 854 – 58
(Reapproved 1979)

American Association State
Highway and Transportation Officials Standard
AASHTO No.: T 100

Standard Test Method for
SPECIFIC GRAVITY OF SOILS[1]

This standard is issued under the fixed designation D 854; the number immediately following the designation indicates the year of original adoption or, in the case of revision, the year of last revision. A number in parentheses indicates the year of last reapproval.

1. Scope

1.1 This method covers determination of the specific gravity of soils by means of a pycnometer. When the soil is composed of particles larger than the No. 4 (4.75-mm) sieve, the method outlined in ASTM Method C 127, Test for Specific Gravity and Absorption of Coarse Aggregate,[2] shall be followed. When the soil is composed of particles both larger and smaller than the No. 4 sieve, the sample shall be separated on the No. 4 sieve and the appropriate test method used on each portion. The specific gravity value for the soil shall be the weighted average of the two values (Note 1). When the specific gravity value is to be used in calculations in connection with the hydrometer portion of ASTM Method D 422, Particle-Size Analysis of Soils,[3] it is intended that the specific gravity test be made on that portion of the soil which passes the No. 10 (2.00-mm) sieve.

NOTE 1—The weighted average specific gravity should be calculated using the following equation:

$$G_{avg} = \cfrac{1}{\cfrac{R_1}{100G_1} + \cfrac{P_1}{100G_2}}$$

where:

G_{avg} = weighted average specific gravity of soils composed of particles larger and smaller than the No. 4 (4.75-mm) sieve,

R_1 = percent of soil particles retained on the No. 4 sieve,

P_1 = percent of soil particles passing the No. 4 sieve,

G_1 = apparent specific gravity of soil particles retained on the No. 4 sieve as determined

by Method C 127, and

G_2 = specific gravity of soil particles passing the No. 4 sieve as determined by this test method.

2. Definition

2.1 *specific gravity*[4]—the ratio of the weight in air of a given volume of a material at a stated temperature to the weight in air of an equal volume of distilled water at a stated temperature.

3. Apparatus

3.1 *Pycnometer*—Either a volumetric flask having a capacity of at least 100 ml or a stoppered bottle having a capacity of at least 50 ml (Note 2). The stopper shall be of the same material as the bottle, and of such size and shape that it can be easily inserted to a fixed depth in the neck of the bottle, and shall have a small hole through its center to permit the emission of air and surplus water.

NOTE 2—The use of either the volumetric flask or the stoppered bottle is a matter of individual preference, but in general, the flask should be used when a larger sample than can be used in the stoppered bottle is needed due to maximum grain size of the sample.

---

[1] This method is under the jurisdiction of ASTM Committee D-18 on Soil and Rock.
Current edition approved Sept. 22, 1958. Originally issued 1945. Replaces D 854 – 55.
[2] *Annual Book of ASTM Standards*, Part 14.
[3] *Annual Book of ASTM Standards*, Part 19.
[4] This definition conforms to ASTM Definitions E 12, Terms Relating to Density and Specific Gravity of Solids, Liquids, and Gases, See *Annual Book of ASTM Standards*, Part 41.

D 854

4. Calibration of Pycnometer

4.1 The pycnometer shall be cleaned, dried, weighed, and the weight recorded. The pycnometer shall be filled with distilled water (Note 3) essentially at room temperature. The weight of the pycnometer and water, W_a, shall be determined and recorded. A thermometer shall be inserted in the water and its temperature T_i determined to the nearest whole degree.

NOTE 3—Kerosine is a better wetting agent than water for most soils and may be used in place of distilled water for oven-dried samples.

4.2 From the weight W_a determined at the observed temperature T_i a table of values of weights W_a shall be prepared for a series of temperatures that are likely to prevail when weights W_b are determined later (Note 4). These values of W_a shall be calculated as follows:

$$W_a \text{ (at } T_x)$$
$$= (\text{density of water at } T_x/\text{density of water at } T_i)$$
$$\times (W_a(\text{at } T_i) - W_f) + W_f$$

where:

W_a = weight of pycnometer and water, g,

W_f = weight of pycnometer, g,

T_i = observed temperature of water, deg C, and

T_x = any other desired temperature, deg C.

NOTE 4—This method provides a procedure that is most convenient for laboratories making many determinations with the same pycnometer. It is equally applicable to a single determination. Bringing the pycnometer and contents to some designated temperature when weights W_a and W_b are taken, requires considerable time. It is much more convenient to prepare a table of weights W_a for various temperatures likely to prevail when weights W_b are taken. It is important that weights W_a and W_b be based on water at the same temperature. Values for the relative density of water at temperatures from 18 to 30 C are given in Table 1.

5. Sample

5.1 The soil to be used in specific gravity test may contain its natural moisture or be oven-dried. The weight of the test sample on an oven-dry basis shall be at least 25 g when the volumetric flask is to be used, and at least 10 g when the stoppered bottle is to be used.

3.2 (continued)

3.2 *Balance*—Either a balance sensitive to 0.01 g for use with the volumetric flask, or a balance sensitive to 0.001 g for use with the stoppered bottle.

5.2 *Samples Containing Natural Moisture*—When the sample contains its natural moisture, the weight of the soil, W_o, on an oven-dry basis shall be determined at the end of the test by evaporating the water in an oven maintained at 230 ± 9 F (110 ± 5 C) (Note 5). Samples of clay soils containing their natural moisture content shall be dispersed in distilled water before placing in the flask, using the dispersing equipment specified in Method D 422 (Note 6).

5.3 *Oven-Dried Samples*—When an oven-dried sample is to be used, the sample shall be dried for at least 12 h, or to constant weight, in an oven maintained at 230 ± 9 F (110 ± 5 C) (Note 5), cooled in a desiccator, and weighed upon removal from the desiccator. The sample shall then be soaked in distilled water for at least 12 h.

NOTE 5—Drying of certain soils at 110 C may bring about loss of moisture of composition or hydration, and in such cases drying shall be done, if desired, in reduced air pressure and at a lower temperature.

NOTE 6—The minimum volume of slurry that can be prepared by the dispersing equipment specified in Method D 422 is such that a 500-ml flask is needed as the pycnometer.

6. Procedure

6.1 Place the sample in the pycnometer, taking care not to lose any of the soil in case the weight of the sample has been determined. Add distilled water to fill the volumetric flask about three-fourths full or the stoppered bottle about half full.

6.2 Remove entrapped air by eight of the following methods: (1) subject the contents to a partial vacuum (air pressure not exceeding 100 mm Hg) or (2) boil gently for at least 10 min while occasionally rolling the pycnometer to assist in the removal of the air. Subject the contents to reduced air pressure either by connecting the pycnometer directly to an aspirator or vacuum pump, or by use of a bell jar. Some soils boil violently when subjected to reduced air pressure. It will be necessary in those cases to reduce the air pressure at a slower rate or to use a larger flask. Cool samples that are heated to room temperature.

6.3 Fill the pycnometer with distilled water, clean the outside and dry with a clean, dry cloth. Determine the weight of the pycnometer and contents, W_b, and the temperature in de-

D 854

grees Celsius, T_x, of the contents as described in Section 4.

7. Calculation and Report

7.1 Calculate the specific gravity of the soil, based on water at a temperature T_x, as follows:

Specific gravity, $T_x/T_x = W_o/[W_o + (W_a - W_b)]$

where:

W_o = weight of sample of oven-dry soil, g,

W_a = weight of pycnometer filled with water at temperature T_x (Note 7), g,

W_b = weight of pycnometer filled with water and soil at temperature T_x, g, and

T_x = temperature of the contents of the pycnometer when weight W_b was determined, deg C.

NOTE 7—This value shall be taken from the table of values of W_a, prepared in accordance with 4.2, for the temperature prevailing when weight W_b was taken.

7.2 Unless otherwise required, specific gravity values reported shall be based on water at 20 C. The value based on water at 20 C shall be calculated from the value based on water at the observed temperature T_x, as follows:

Specific gravity, $T_x/20$ C =
$$K \times \text{specific gravity, } T_x/T_x$$

where:

K = a number found by dividing the relative density of water at temperature T_x by the relative density of water at 20 C. Values for a range of temperatures are given in Table 1.

7.3 When it is desired to report the specific gravity value based on water at 4 C, such a specific gravity value may be calculated by multiplying the specific gravity value at temperature T_x by the relative density of water at temperature T_x.

7.4 When any portion of the original sample of soil is eliminated in the preparation of the test sample, the portion on which the test has been made shall be reported.

TABLE 1   Relative Density of Water and Conversion Factor K For Various Temperatures

Temperature, deg C	Relative Density of Water	Correction Factor K
18	0.9986244	1.0004
19	0.9984347	1.0002
20	0.9982343	1.0000
21	0.9980233	0.9998
22	0.9978019	0.9996
23	0.9975702	0.9993
24	0.9973286	0.9991
25	0.9970770	0.9989
26	0.9968156	0.9986
27	0.9965451	0.9983
28	0.9962652	0.9980
29	0.9959761	0.9977
30	0.9956780	0.9974

ASTM D 1140 – 54 (Reapproved 1971)

Standard Test Method for

AMOUNT OF MATERIAL IN SOILS FINER THAN THE NO. 200 (75-μm) SIEVE[1]

This standard is issued under the fixed designation D 1140; the number immediately following the designation indicates the year of original adoption or, in the case of revision, the year of last revision. A number in parentheses indicates the year of last reapproval.

This method has been approved for use by agencies of the Department of Defense and for listing in the DoD Index of Specifications and Standards.

1. Scope

1.1 This method covers determination of the total amount of material in soils finer than the No. 200 (75-μm) sieve.

2. Apparatus

2.1 *Sieves*—A nest of two sieves, the lower being a No. 200 (75-μm) sieve and the upper a No. 40 (425-μm) sieve, both conforming to ASTM Specification E 11, for Wire-Cloth Sieves for Testing Purposes.[2]

2.2 *Containers*—A pan or vessel of sufficient size to contain the test sample covered with water and to permit of vigorous agitation without advertent loss of any part of the sample, and a second pan or container for use in drying the test sample after washing.

3. Test Sample

3.1 The test sample shall be selected from material that has been thoroughly mixed. A representative sample, sufficient to yield not less than the approximate weight of dried material shown in the following table, shall be selected using a sample splitter or by the method of quartering:

Nominal Diameter of Largest Particle, in.	Approximate Minimum Weight of Sample, g
0.0787 (No. 10 sieve) (2.0 mm)	200
0.187 (No. 4 sieve) (4.75 mm)	500
¾ (19.0 mm)	1500
1 (25.0 mm)	2000
1½ or over (37.5 mm)	2500

4. Procedure

4.1 Dry the test sample to a constant weight at a temperature not exceeding 230 ± 9 F (110 ± 5 C) and weigh to the nearest 0.05 percent, or alternatively, weigh the test sample moist and use an auxiliary moisture content sample to determine the moisture content of the sample. The weight of the moisture content sample shall be between 20 and 30 percent of the weight of the test sample. Calculate the oven-dry weight of the test sample from the moist weight and the moisture content.

4.2 Place the test sample in the container, add sufficient clean water to cover it, and allow to soak a minimum of 2 h (preferably overnight).

4.3 Agitate the contents of the container vigorously and pour the wash water immediately over the nested sieves, arranged with the coarser sieve on top. Repeat the process of adding clear water to the container to cover the sample, agitating the contents of the container, and pouring the wash water over the nested sieves until the wash water is clear. When the total sample is small, the entire contents of the soaking container may be transferred to the nested sieves after the first washing and the

---

[1] This method is under the jurisdiction of ASTM Committee D-18 on Soil and Rock.
Current edition approved Sept. 15, 1954. Originally issued 1950. Replaces D 1140 – 50 T.
[2] *Annual Book of ASTM Standards*, Part 41.

ASTM **D 1140**

washing operation completed in accordance with 4.4. The wash water need not be saved.

NOTE 1—The percentage value secured at the end of the test may not be correct (being too low) for soils containing relatively high percentages of the minus 200 fraction. This appears to be due chiefly to inadequate agitation. When it is desired to secure the exact percentage for the minus 200 fraction for such a soil, the portion of the sample passing the No. 40 sieve and retained on the No. 200 sieve secured in the washing operation, shall be transferred to the dispersion cup of the stirring apparatus used in Method D 422, Particle-Size Analysis of Soils,[3] the cup filled half full with water and the contents agitated for 1 min. After this agitation the contents of the cup shall be transferred to the nested sieves and washing continued.

If the stirring apparatus has not been used prior to the drying of the portion of the sample larger than the No. 200 (75-μm) sieve, and it is desired to do so after drying, the dried material shall be separated on the No. 40 (425-μm) sieve; the portion retained shall be saved; and the portion passing shall be placed in the dispersion cup with water and agitated for 1 min with the stirring apparatus as previously described. The contents of the cup shall be transferred to the No. 200 sieve, washed, and dried. The revised total weight retained on the No. 200 sieve shall be secured by combining and weighing the two fractions.

4.4 Transfer the sample to the nested sieves and wash with running water (Note 2). When the sample is larger than can be handled at one time on the nested sieves, wash a portion of the sample and transfer to the container in which it is to be dried.

NOTE 2—Tapping of sieves has been found to expedite the washing operations.

4.5 Dry the washed material retained on the nested sieves in a container to a constant weight at a temperature not exceeding 230 ± 9 F (110 ± 5 C) and dry-sieve it on the nested sieves (Note 3). Weigh the dry material retained on the nested sieves to the nearest 0.05 percent.

NOTE 3—Some material passes the No. 200 (75-μm) sieve on dry sieving that did not pass during the washing operation. When desired, a sieve analysis may be made on the portion of the sample retained on the No. 200 sieve, in accordance with Method D 422.

5. Calculation

5.1 Calculate the results as follows:

$$P = [(W_o - W_1)/W_o] \times 100$$

where:

P = percentage of material finer than No. 200 (75-μm) sieve,

W_o = weight of original sample on an oven-dry basis, g and

W_1 = oven-dry weight of sample after washing and dry-sieving, g

---

[3] *Annual Book of ASTM Standards*, Part 19.

ASTM D 1556 – 64 (Reapproved 1974)[ε1]

Standard Test Method for

DENSITY OF SOIL IN PLACE BY THE SAND-CONE METHOD[1]

This standard is issued under the fixed designation D 1556; the number immediately following the designation indicates the year of original adoption or, in the case of revision, the year of last revision. A number in parentheses indicates the year of last reapproval.

[ε1] NOTE—The equation in 4.4 was editorially corrected in July 1974.

1. Scope

1.1 This method covers the determination of the in-place density of soils. The apparatus described herein is restricted to tests in soils containing particles not larger than 2 in. (50.8 mm) in diameter.

2. Apparatus

2.1 *Density Apparatus*—The density apparatus shall consist of a 1-gal (4-litre) jar and a detachable appliance consisting of a cylindrical valve with an orifice ½ in. (12.7 mm) in diameter and having a small funnel continuing to a standard *G* mason jar top on one end and a large funnel on the other end. The valve shall have stops to prevent rotating the valve past the completely open or completely closed positions. The apparatus shall conform to the requirements shown in Fig. 1.

NOTE 1—The apparatus described here represents a design that has proved satisfactory. Other apparatus of similar proportions will perform equally well so long as the basic principles of the sand-volume determination are observed. This apparatus, when full, can be used with test holes having a volume of approximately 0.1 ft³ (3 dm³). The base plate shown in the drawing is optional; its use may make leveling more difficult but permits test holes of larger diameter and may reduce loss in transferring soil from test-hole to container as well as afford a more constant base for tests in soft soils. When the base plate is used it shall be considered a part of the funnel in the procedures of this test method.

2.2 *Sand*—Any clean, dry, free-flowing, uncemented sand having few, if any, particles passing the No. 200 (75-μm) or retained on the No. 10 (2.00-mm) sieves. In selecting a sand for use several bulk density determinations should be made using the same representative sample for each determination. To be acceptable the sand shall not have a variation in bulk

density greater than 1 percent.

2.3 *Balances*—A balance or scale of 10-kg capacity accurate to 1.0 g and a balance of 500-g capacity accurate to 0.1 g.

2.4 *Drying Equipment*—Stove or oven or other suitable equipment for drying moisture content samples.

2.5 *Miscellaneous Equipment*—Small pick, chisels, or spoons for digging test hole; 10-in. (254-mm) frying pan or any suitable container for drying moisture samples; buckets with lids, seamless tin cans with lids, canvas sacks or other suitable containers for retaining the density sample, moisture sample or density sand respectively; thermometer for determining the temperature of water, small paint-type brush, slide rule, note-book, etc.

3. Procedure

3.1 Determine the volume of the jar and attachment up to and including the volume of the valve orifice as follows (Note 2):

3.1.1 Weigh the assembled apparatus and record.

3.1.2 Place the apparatus upright and open the valve.

3.1.3 Fill the apparatus with water until it appears over the valve.

3.1.4 Close valve and remove excess water.

3.1.5 Weigh the apparatus and water and determine the temperature of the water.

3.1.6 Repeat the procedure described 3.1.1 to 3.1.5 at least twice. Convert the weight of water, in grams, to millilitres by correcting for the temperature as given in 4.1. The volume

---

[1] This method is under the jurisdiction of ASTM Committee D-18 on Soil and Rock.
Current edition approved Aug. 31, 1964. Originally issued 1958. Replaces D 1556 – 58 T.

D 1556

used shall be the average of three determinations with a maximum variation of 3 ml.

Note 2—The volume determined in this procedure is constant as long as the jar and attachment are in the same relative position. If the two are to be separated match marks should be made to permit reassembly to this position.

3.2 Determine the bulk density of the sand to be used in the field test as follows (Notes 3 and 4):

3.2.1 Place the empty apparatus upright on a firm level surface, close the valve, and fill the funnel with sand.

3.2.2 Open the valve and, keeping the funnel at least half full of sand, fill the apparatus. Close the valve sharply and empty excess sand.

3.2.3 Weigh the apparatus with sand and determine the net weight of sand by subtracting the weight of the apparatus.

Note 3—Vibration of the sand during any sand weight-volume determination may increase the bulk density of the sand and decrease the accuracy of the determination. Appreciable time intervals between the bulk density determination of the sand and its use in the field may result in change in the bulk density caused by a change in the moisture content or effective gradation.

Note 4—It is possible to determine the bulk density of the sand in other containers of known volume that dimensionally approximate the largest test hole that will be dug. The general procedure used is that given in 3.4 for determining the volume of the test hole. If this procedure is to be followed it shall be determined that the resulting bulk density equals that given by the jar determination.

3.3 Determine the weight of sand required to fill the funnel as follows (Notes 5 and 6):

3.3.1 Put sand in the apparatus and secure the weight of apparatus and sand.

3.3.2 Seat the inverted apparatus on a clean, level, plane surface.

3.3.3 Open the valve and keep open until after the sand stops running.

3.3.4 Close the valve sharply, weigh the apparatus with remaining sand, and determine the loss of sand. This loss represents the weight of sand required to fill the funnel.

3.3.5 Replace the sand removed in the funnel determination and close the valve.

Note 5—This determination may be omitted if the procedure given in Note 7 is followed. When the base plate is used it shall be considered a part of the funnel.

Note 6—Where test holes of maximum volume are desired it is possible, after the bulk density determination, to settle the sand by vibration and increase the weight of sand in the apparatus. If this procedure is followed, the total weight of sand

available shall be determined by reweighing.

3.4 Determine the density of the soil in place as follows:

3.4.1 Prepare the surface of the location to be tested so that it is a level plane.

3.4.2 Seat the inverted apparatus on the prepared plane surface and mark the outline of the funnel.

Note 7—In soils such that leveling is not successful a preliminary test shall be run at this point measuring the volume bounded by the funnel and ground surface. This step requires balances at the test site or emptying and refilling the apparatus. After this measurement is completed, carefully brush the sand from the prepared surface.

3.4.3 Dig the test hole inside the funnel mark, being very careful to avoid disturbing the soil that will bound the hole. Soils that are essentially granular require extreme care. Place all loosened soil in a container, being careful to avoid losing any material.

3.4.4 Seat the apparatus in the previously marked position, open the valve, and after the sand has stopped flowing, close the valve (Note 3).

3.4.5 Weigh the apparatus with remaining sand, and determine the weight of sand used in the test.

3.4.6 Weigh the material that was removed from the test hole.

3.4.7 Mix the material thoroughly and secure and weigh a representative sample for moisture determination.

3.4.8 Dry and weigh the moisture sample.

3.5 The minimum test hole volumes suggested in determining the in-place density of soil mixtures are given in Table 1. This table shows the suggested minimum weight of the moisture content sample in relation to the maximum particle size in soil mixtures.

4. Calculations

4.1 Calculate the volume of the density apparatus as follows:

$$V_1 = GT$$

where:

V_1 = volume of the density apparatus, ml,

G = weight of water required to fill the apparatus, g, and

T = water temperature-volume correction shown in column 3 of Table 2.

4.2 Calculate the bulk density of the sand as follows:

D 1556

$$W_1 = 62.427 \ W_2/V_1$$

where:

W_1 = bulk density of the sand, lb/ft^3,

W_2 = weight of sand required to fill the apparatus (3.2.3), g, and

V_1 = volume of apparatus (4.1).

4.3 Calculate the moisture content and the dry weight of material removed from the test hole as follows:

$$w = [(W_3 - W_4)/W_4] \times 100$$
$$W_6 = 0.2205 \ W_5/(w + 100)$$

where:

w = percentage of moisture, in material from test hole,

W_3 = moist weight of moisture sample, g,

W_4 = dry weight of moisture sample, g,

W_5 = moist weight of the material from the test hole, g, and

W_6 = dry weight of material from test hole, lb.

4.4 Calculate the in-place dry density of the material tested as follows:

$$V = (W_7 - W_8)/453.6 \ W_1$$
$$W = W_6/V$$

where:

V = volume of test hole, ft^3,

W_7 = weight of sand used (3.4.5), g,

W_8 = weight of sand in funnel (3.3.3), g, and

W = dry density of the tested material, lb/ft^3.

NOTE 8—It may be desired to express the in-place density as a percentage of some other density, for example, the laboratory maximum density determined in accordance with Method D 698, Test for Moisture-Density Relations of Soils, Using a 5.5-lb (2.49-kg) Rammer and a 12-in. (305-mm) Drop.[2] This relation can be determined by dividing the in-place density by the maximum density and multiplying by 100.

---

[2] *Annual Book of ASTM Standards*, Part 19.

TABLE 1 Minimum Test Hole Volumes and Minimum Moisture Content Samples Based on Maximum Size of Particle

Maximum Particle Size	Minimum Test Hole Volume, ft^3	Minimum Moisture Content Sample, g
No. 4 Sieve (4.75 mm)	0.025	100
½ in. (12.5 mm)	0.050	250
1 in. (25 mm)	0.075	500
2 in. (50 mm)	0.100	1000

TABLE 2 Volume of Water per Gram Based on Temperature

Temperature		Volume of Water, ml/g
deg C	deg F	
12	53.6	1.00048
14	57.2	1.00073
16	60.8	1.00103
18	64.4	1.00138
20	68.0	1.00177
22	71.6	1.00221
24	75.2	1.00268
26	78.8	1.00320
28	82.4	1.00375
30	86.0	1.00435
32	89.6	1.00497

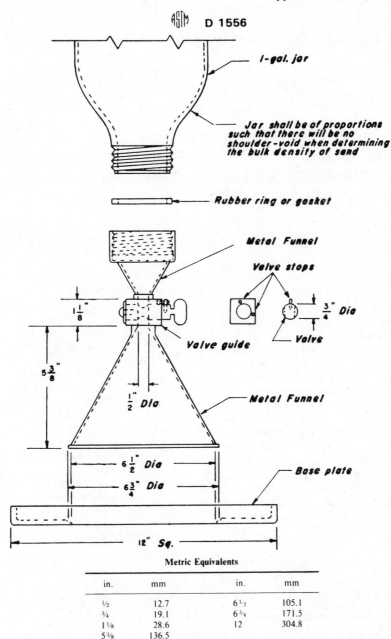

ⒶⓈⓉⓂ   **D 1556**

1-gal. jar

*Jar shall be of proportions
such that there will be no
shoulder-void when determining
the bulk density of sand*

Rubber ring or gasket

Metal Funnel

Valve stops

$1\frac{1}{8}$"

$\frac{3}{4}$" *Dia*

Valve

Valve guide

$5\frac{3}{8}$"

$\frac{1}{2}$" *Dia*

Metal Funnel

$6\frac{1}{2}$" *Dia*

Base plate

$6\frac{3}{4}$" *Dia*

12" *Sq.*

Metric Equivalents

in.	mm	in.	mm
$\frac{1}{2}$	12.7	$6\frac{1}{2}$	105.1
$\frac{3}{4}$	19.1	$6\frac{3}{4}$	171.5
$1\frac{1}{8}$	28.6	12	304.8
$5\frac{3}{8}$	136.5		

FIG. 1 Density Apparatus.

The American Society for Testing and Materials takes no position respecting the validity of any patent rights asserted in connection with any item mentioned in this standard. Users of this standard are expressly advised that determination of the validity of any such patent rights, and the risk of infringement of such rights, is entirely their own responsibility.

This standard is subject to revision at any time by the responsible technical committee and must be reviewed every five years and if not revised, either reapproved or withdrawn. Your comments are invited either for revision of this standard or for additional standards and should be addressed to ASTM Headquarters. Your comments will receive careful consideration at a meeting of the responsible technical committee, which you may attend. If you feel that your comments have not received a fair hearing you should make your views known to the ASTM Committee on Standards, 1916 Race St., Philadelphia, Pa. 19103, which will schedule a further hearing regarding your comments. Failing satisfaction there, you may appeal to the ASTM Board of Directors.

ASTM D 1557 – 78

Standard Test Methods for
MOISTURE-DENSITY RELATIONS OF SOILS AND SOIL-AGGREGATE MIXTURES USING 10-lb (4.54-kg) RAMMER AND 18-in. (457-mm) DROP[1]

This standard is issued under the fixed designation D 1557; the number immediately following the designation indicates the year of original adoption or, in the case of revision, the year of last revision. A number in parentheses indicates the year of last reapproval.

1. Scope

1.1 These laboratory compaction methods cover the determination of the relationship between the moisture content and density of soils and soil-aggregate mixtures (Note 1) when compacted in a mold of a given size with a 10-lb (4.54-kg) rammer dropped from a height of 18 in. (457 mm) (Note 2). Four alternative procedures are provided as follows:

1.1.1 *Method A*—A 4-in. (101.6-mm) mold; material passing a No. 4 (4.75-mm) sieve;

1.1.2 *Method B*—A 6-in. (152.4-mm) mold; material passing a No. 4 (4.75-mm) sieve;

1.1.3 *Method C*—A 6-in. (152.4-mm) mold; material passing a ¾-in. (19.0-mm) sieve; and

1.1.4 *Method D*—A 6-in. (152.4-mm) mold; material passing a ¾-in. (19.0-mm) sieve, corrected by replacement for material retained on a ¾-in. sieve.

NOTE 1—Soils and soil-aggregate mixtures should be regarded as natural occurring fine- or coarse-grained soils or composites or mixtures of natural soils, or mixtures of natural and processed soils or aggregates such as silt, gravel, or crushed rock.

NOTE 2—These laboratory compaction test methods when used on soils and soil-aggregates which are not free-draining will, in most cases, establish a well-defined optimum moisture content and maximum density (see Section 7). However, for free-draining soils and soil-aggregate mixtures, these methods will not, in many cases, produce a well-defined moisture-density relationship and the maximum density obtained will generally be less than that obtained by vibratory methods.

1.2 The method to be used should be indicated in the specifications for the material being tested. If no method is specified, the provisions of Section 5 shall govern.

2. Applicable Documents

2.1 *ASTM Standards*:

C 127 Test for Specific Gravity and Absorption of Coarse Aggregate[2]

D 854 Test for Specific Gravity of Soils[3]

D 2168 Calibration of Laboratory Mechanical-Rammer Soil Compactors[3]

D 2216 Laboratory Determination of Water (Moisture) Content of Soil, Rock, and Soil-Aggregate Mixtures[3]

D 2487 Classification of Soils for Engineering Purposes[3]

D 2488 Recommended Practice for Description of Soils (Visual-Manual Procedure)[3]

E 11 Specification for Wire-Cloth Sieves for Testing Purposes[4]

3. Apparatus

3.1 *Molds*—The molds shall be cylindrical in shape, made of rigid metal and be within the capacity and dimensions indicated in 3.1.1 or 3.1.2. The molds may be the "split" type, consisting either of two half-round sections, or a section of pipe split along one element,

[1] These methods are under the jurisdiction of ASTM Committee D-18 on Soil and Rock.
Current edition approved April 27, 1978. Published July 1978. Originally published as D 1557 – 58 T. Last previous edition D 1557 – 70.
[2] *Annual Book of ASTM Standards*, Parts 14 and 15.
[3] *Annual Book of ASTM Standards*, Part 19.
[4] *Annual Book of ASTM Standards*, Parts 14, 15, and 41.

ASTM **D 1557**

which can be securely locked together to form a cylinder meeting the requirements of this section. The molds may also be the "taper" type, providing the internal diameter taper is uniform and is not more than 0.200 in./linear ft (16.7 mm/linear m) of mold height. Each mold shall have a base plate assembly and an extension collar assembly, both made of rigid metal and constructed so they can be securely attached to or detached from the mold. The extension collar assembly shall have a height extending above the top of the mold of at least 2 in. (50 mm) which may include an upper section that flares out to form a funnel providing there is at least a ¾-in. (19-mm) straight cylindrical section beneath it.

3.1.1 *Mold*, 4.0 in. (101.6 mm) in diameter, having a capacity of $^1/_{30}$ ± 0.0004 ft³ (944 ± 11 cm³) and conforming to Fig. 1.

3.1.2 *Mold*, 6.0 in. (152.4 mm) in diameter, having a capacity of $^1/_{13.333}$ ± 0.0009 ft³ (2124 ± 25 cm³) and conforming to Fig. 2.

3.1.3 The average internal diameter, height, and volume of each mold shall be determined before initial use and at intervals not exceeding 1000 times the mold is filled. The mold volume shall be calculated from the average of at least six internal diameter and three height measurements made to the nearest 0.001 in. (0.02 mm), or from the amount of water required to completely fill the mold, corrected for temperature variance in accordance with Table 1. If the average internal diameter and volume are not within the tolerances shown in Figs. 1 or 2, the mold shall not be used. The determined volume shall be used in computing the required densities.

3.2 *Rammer* — The rammer may be either manually operated (see 3.2.1) or mechanically operated (see 3.2.2). The rammer shall fall freely through a distance of 18.0 ± $^1/_{16}$ in. (457.2 ± 1.6 mm) from the surface of the specimen. The manufactured weight of the rammer shall be 10.00 ± 0.02 lb (4.54 ± 0.01 kg). The specimen contact face shall be flat.

3.2.1 *Manual Rammer* — The specimen contact face shall be circular with a diameter of 2.000 ± 0.005 in. (50.80 ± 0.13 mm). The rammer shall be equipped with a guide-sleeve which shall provide sufficient clearance so that the free fall of the rammer shaft and head will not be restricted. The guidesleeve

shall have four vent holes at each end (eight holes total) located with centers ¾ ± $^1/_{16}$ in. (19.0 ± 1.6 mm) from each end and spaced 90 deg apart. The minimum diameter of the vent holes shall be ³/₈ in. (9.5 mm).

3.2.2 *Mechanical Rammer* — The rammer shall operate mechanically in such a manner as to provide uniform and complete coverage of the specimen surface. There shall be 0.10 ± 0.03 in. (2.5 ± 0.8 mm) clearance between the rammer and the inside surface of the mold at its smallest diameter. When used with the 4.0-in. (101.6-mm) mold, the specimen contact face shall be circular with a diameter of 2.000 ± 0.005 in. (50.80 ± 0.13 mm). When used with the 6.0-in. (152.4-mm) mold, the specimen contact face shall have the shape of a section of a circle of a radius equal to 2.90 ± 0.02 in. (73.7 ± 0.5 mm). The sector face rammer shall operate in such a manner that the vertex of the sector is positioned at the center of the specimen. The mechanical rammer shall be calibrated and adjusted, as necessary, in accordance with 3.2.3.

3.2.3 *Calibration and Adjustment* — The mechanical rammer shall be calibrated, and adjusted as necessary, before initial use; near the end of each period during which the mold was filled 1000 times; before reuse after anything, including repairs, which may affect the test results significantly; and whenever the test results are questionable. Each calibration and adjustment shall be in accordance with Method D 2168.

3.3 *Sample Extruder* (optional)—A jack, frame, or other device adapted for the purpose of extruding compacted specimens from the mold.

3.4 *Balances* — A balance or scale of at least 20-kg capacity sensitive to ±1 g and a balance of at least 1000-g capacity sensitive to ±0.01 g.

3.5 *Drying Oven*, thermostatically-controlled, preferably of the forced-draft type, capable of maintaining a temperature of 230 ± 9°F (110 ± 5°C) for determining the moisture content of the compacted specimen.

3.6 *Straightedge* — A stiff metal straightedge of any convenient length but not less than 10 in. (254 mm). The scraping edge shall have a straightness tolerance of ±0.005 in. (±0.13 mm) and shall be beveled if it is thicker then ⅛ in. (3 mm).

3.7 *Sieves*, 3-in. (75-mm), ¾-in. (19.0-mm), and No. 4 (4.75-mm), conforming to the requirements of Specification E 11.

3.8 *Mixing Tools* – Miscellaneous tools such as mixing pan, spoon, trowel, spatula, etc., or a suitable mechanical device for thoroughly mixing the sample of soil with increments of water.

4. Procedure

4.1 *Specimen Preparation* – Select a representative portion of quantity adequate to provide, after sieving, an amount of material weighing as follows: Methods A – 25 lb (11 kg); Methods B, C, and D – 50 lb (23 kg). Prepare specimens in accordance with either 4.1.1 through 4.1.3 or 4.1.4.

4.1.1 *Dry Preparation Procedure* – If the sample is too damp to be friable, reduce the moisture content by drying until the material is friable; see 4.1.2. Drying may be in air or by the use of a drying apparatus such that the temperature of the sample does not exceed 140°F (60°C). After drying (if required), thoroughly break up the aggregations in such a manner as to avoid reducing the natural size of the particles. Pass the material through the specified sieve as follows: Methods A and B – No. 4 (4.75-mm); Methods C and D – ¾-in. (19.0-mm). Correct for oversize material in accordance with Section 5, if Method D is specified.

4.1.2 Whenever practicable, soils classified as ML, CL, OL, GC, SC, MH, CH, OH and PT by Classification D 2487 shall be prepared in accordance with 4.1.4.

4.1.3 Prepare a series of at least four specimens by adding increasing amounts of water to each sample so that the moisture contents vary by approximately 1½ %. The moisture contents selected shall bracket the optimum moisture content, thus providing specimens which, when compacted, will increase in mass to the maximum density and then decrease in density (see 7.2 and 7.3). Thoroughly mix each specimen to ensure even distribution of moisture throughout and then place in a separate covered container and allow to stand prior to compaction in accordance with Table 2. For the purpose of selecting a standing time, it is not required to perform the actual classification procedures described in Classification D 2487 (except in the case of referee testing), if

previous data exist which provide a basis for classifying the sample.

4.1.4 *Moist Preparation Method* – The following alternate procedure is recommended for soils classified as ML, CL, OL, GC, SC, MH, CH, OH and PT by Classification D 2487. Without previously drying the sample, pass it through the ¾-in. (19.0-mm) and No. 4 (4.75-mm) sieves. Correct for oversize material in accordance with Section 5, if Method D is specified. Prepare a series of at least four specimens having moisture contents that vary by approximately 1½ %. The moisture contents selected shall bracket the optimum moisture content, thus providing specimens which, when compacted, will increase in mass to the maximum density and then decrease in density (see 7.2 and 7.3). To obtain the appropriate moisture content of each specimen, the addition of a predetermined amount of water (see 4.1.3) or the removal of a predetermined amount of moisture by drying may be necessary. Drying may be in air or by the use of a drying apparatus such that the temperature of the specimen does not exceed 140°F (60°C). The prepared specimens shall then be thoroughly mixed and stand, as specified in 4.1.3 and Table 2, prior to compaction.

NOTE 3 – With practice, it is usually possible to visually judge the point of optimum moisture closely enough so that the prepared specimens will bracket the point of optimum moisture content.

4.2 *Specimen Compaction* – Select the proper compaction mold, in accordance with the method being used, and attach the mold extension collar. Compact each specimen in five layers of approximately equal height. Each layer shall receive 25 blows in the case of the 4-in. (101.6-mm) mold; each layer shall receive 56 blows in the case of the 6-in. (152.4-mm) mold. The total amount of material used shall be such that the fifth compacted layer is slightly above the top of the mold, but not exceeding ¼ in. (6 mm). During compaction the mold shall rest on a uniform rigid foundation, such as provided by a cylinder or cube of concrete weighing not less than 200 lb (91 kg).

4.2.1 In operating the manual rammer, care shall be taken to avoid rebound of the rammer from the top end of the guidesleeve.

ASTM **D 1557**

The guidesleeve shall be held steady and within 5 deg of the vertical. The blows shall be applied at a uniform rate not exceeding approximately 1.4 s per blow and in such a manner as to provide complete and uniform coverage of the specimen surface.

4.2.2 *Mold Sizes*—The mold size used shall be as follows: Method A, 4-in. (101.6-mm); Methods B, C, and D, 6-in. (152.4-mm).

4.2.3 Following compaction, remove the extension collar; carefully trim the compacted specimen even with the top of the mold by means of the straightedge and determine the mass of the specimen. Divide the mass of the compacted specimen and mold, minus the mass of the mold, by the volume of the mold (see 3.1.3). Record the result as the wet density, γ_m, in pounds per cubic foot (or kilograms per cubic metre) of the compacted specimen.

4.2.4 Remove the material from the mold. Determine moisture content in accordance with Method D 2216, using either the whole compacted specimen or alternatively a representative specimen of the whole specimen. The whole specimen must be used when the permeability of the compacted specimen is high enough so that the moisture content is not distributed uniformly throughout. If the whole specimen is used, break it up to facilitate drying. Obtain the representative specimen by slicing the compacted specimen axially through the center and removing 100 to 500 g of material from one of the cut faces.

4.2.5 Repeat 4.2 through 4.2.4 for each specimen prepared.

5. Oversize Corrections

5.1 If 30 % or more of the sample is retained on a ³/₄-in. (19.0-mm) sieve, then none of the methods described under these methods shall be used for the determination of either maximum density or optimum moisture content.

5.2 *Methods A and B*—The material retained on the No. 4 (4.75-mm) sieve is discarded and no oversize correction is made. However, it is recommended that if the amount of material retained is 7 % or greater, Method C be used instead.

5.3 *Method C*—The material retained on the ³/₄-in. (19.0-mm) sieve is discarded and no oversize correction is made. However, if

the amount of material retained is 10 % or greater, it is recommended that Method D be used instead.

5.4 *Method D:*

5.4.1 This method shall not be used unless the amount of material retained on the ³/₄-in. (19.0-mm) sieve is 10 % or greater. When the amount of material retained on the ³/₄-in. sieve is less than 10 %, use Method C.

5.4.2 Pass the material retained on the ³/₄-in. (19.0-mm) sieve through a 3-in. or 75-mm sieve. Discard the material retained on the 3-in. sieve. The material passing the 3-in. sieve and retained on the ³/₄-in. sieve shall be replaced with an equal amount of material passing a ³/₄-in. sieve and retained on a No. 4 (4.75-mm) sieve. The material for replacement shall be taken from an unused portion of the sample.

6. Calculations

6.1 Calculate the moisture content and the dry density of each compacted specimen as follows:

$$w = [(A - B)/(B - C)] \times 100$$

and

$$\gamma_d = [\gamma_m/(w + 100)] \times 100$$

where:

w = moisture content in percent of the compacted specimens,

A = mass of contained and moist specimen,

B = mass of container and oven-dried specimen,

C = mass of container,

γ_d = dry density, in pounds per cubic foot (or kilograms per cubic metre) of the compacted specimen, and

γ_m = wet density, in pounds per cubic foot (or kilograms per cubic metre) of the compacted specimen.

7. Moisture-Density Relationship

7.1 From the data obtained in 6.1, plot the dry density values as ordinates with corresponding moisture contents as abscissas. Draw a smooth curve connecting the plotted points. Also draw a curve termed the "curve of complete saturation" or "zero air voids curve" on this plot. This curve represents the relationship between dry density and corresponding moisture contents when the voids are completely filled with water. Values of dry density and

D 1557

corresponding moisture contents for plotting the curve of complete saturation can be computed using the following equation:

$$w_{sat} = [(62.4/\gamma_d) - (1/G_s)] \times 100$$

where:

w_{sat} = moisture content in percent for complete saturation,

γ_d = dry density in pounds per cubic foot (or kilograms per cubic metre),

G_s = specific gravity of the material being tested (see Note 4), and

62.4 = density of water in pounds per cubic foot (or kilograms per cubic metre).

NOTE 4 — The specific gravity of the material can either be assumed or based on the weighted average values of: (*a*) the specific gravity of the material passing the No. 4 (4.75-mm) sieve in accordance with Method D 854; and (*b*) the apparent specific gravity of the material retained on the No. 4 (4.75-mm) sieve in accordance with Method C 127.

7.2 *Optimum Moisture Content, w_o* — The moisture content corresponding to the peak of the curve drawn as directed in 7.1 shall be termed the "optimum moisture content."

7.3 *Maximum Density, γ_{max}* — The dry density in pounds per cubic foot (or kilograms per cubic metre) of the sample at "optimum moisture content" shall be termed "maximum density."

8. Report

8.1 The report shall include the following:

8.1.1 Method used (Method A, B, C, or D).

8.1.2 Optimum moisture content.

8.1.3 Maximum density.

8.1.4 Description of rammer (whether manual or mechanical).

8.1.5 Description of appearance of material used in test, based on Recommended Practice D 2488 (Classification D 2487 may be used as an alternative).

8.1.6 Origin of material used in test.

8.1.7 Preparation procedure used (moist or dry).

9. Precision

9.1 Criteria for judging the acceptability of the maximum density and optimum moisture content test results are given in Table 3. The standard deviation s is calculated from the equation:

$$s^2 = \frac{1}{n-1} \sum_1^n (x - \bar{x})^2$$

where:

n = number of determinations;

x = individual value of each determination; and

$\bar{x}$ = numerical average of the determinations.

9.2 Criteria for assigning standard deviation values for single-operator precision are not available at the present time.

TABLE 1 Volume of Water per Gram based on Temperature[A]

Temperature, °C (°F)	Volume of Water, ml/g
12 (53.6)	1.00048
14 (57.2)	1.00073
16 (60.8)	1.00103
18 (64.4)	1.00138
20 (68.0)	1.00177
22 (71.6)	1.00221
24 (75.2)	1.00268
26 (78.8)	1.00320
28 (82.4)	1.00375
30 (86.0)	1.00435
32 (89.6)	1.00497

[A] Values other than shown may be obtained by referring to the *Handbook of Chemistry and Physics,* Chemical Rubber Publishing Co., Cleveland, Ohio.

TABLE 2 Dry Preparation Method — Standing Times

Classification D 2487	Minimum Standing Time, h
GW, GP, SW, SP	no requirement
GM, SM	3
ML, CL, OL, GC, SC	18
MH, CH, OH, PT	36

D 1557

TABLE 3   Precision

	Standard Deviation, s	Acceptable Range of Two Results, Expressed as Percent of Mean Value[A]
Single-operator precision:		
Maximum density	...	1.9
Optimum moisture content	...	9.5
Multilaboratory precision:		
Maximum density	±1.66	4.0
Optimum moisture content	±0.86	15.0

[A] This column indicates a limiting range of values which should not be exceeded by the difference between any two results, expressed as a percentage of the average value. In cooperative test programs it has been determined that 95 % of the tests do not exceed the limiting acceptable ranges shown below. All values shown in this table are based on average test results from a variety of different soils and are subject to future revision.

TABLE 4   Metric Equivalents for Figs. 1 and 2

in.	mm
0.016	0.41
0.026	0.66
1/32	0.8
1/16	1.6
1/8	3.2
1/4	6.4
11/32	8.7
3/8	9.5
1/2	12.7
5/8	15.9
2	50.8
2 1/2	63.5
4	101.6
4 1/4	108.0
4 1/2	114.3
4.584	116.43
6	152.4
6 1/2	165.1
8	203.2

ft³	cm³
1/30	944
0.004	11
1/13,333	2124
0.0009	25

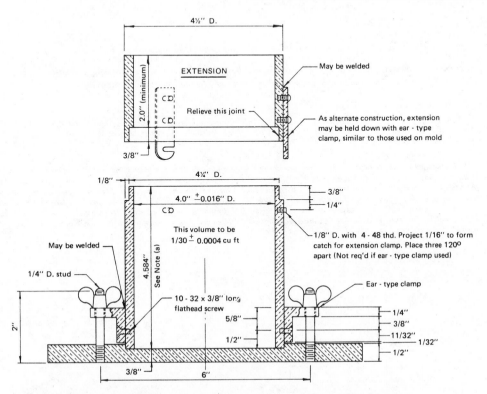

NOTE 1—The tolerance on the height is governed by the allowable volume and diameter tolerances.
NOTE 2—The methods shown for attaching the extension collar to the mold and the mold to the base plate are recommended. However, other methods are acceptable, providing the attachments are equally as rigid as those shown.

FIG. 1   Cylindrical Mold, 4.0-in. for Soil Tests (see Table 4 for metric equivalents).

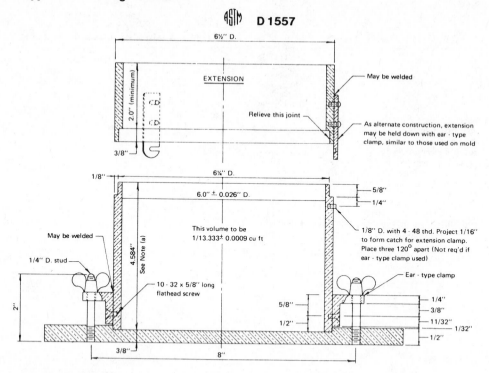

ASTM   D 1557

NOTE 1 – The tolerance on the height is governed by the allowable volume and diameter tolerances.

NOTE 2 – The methods shown for attaching the extension collar to the mold and the mold to the base plate are recommended. However, other methods are acceptable, providing the attachments are equally as rigid as those shown.

FIG. 2   Cylindrical Mold, 6.0-in. for Soil Tests (see Table 4 for metric equivalents).

ASTM D 2049 – 69[ε1]

Standard Test Method for
RELATIVE DENSITY OF COHESIONLESS SOILS[1]

This standard is issued under the fixed designation D 2049; the number immediately following the designation indicates the year of original adoption or, in the case of revision, the year of last revision. A number in parentheses indicates the year of last reapproval.

[ε1] NOTE—Paragraphs 6.1.2 and 8.1 were corrected editorially in December 1977.

1. Scope

1.1 This method[2] covers the determination of the relative density of cohesionless free-draining soils for which impact compaction will not produce a well-defined moisture-density relationship curve and the maximum density by impact methods will generally be less than by vibratory methods. Soils for which this method is applicable may contain up to 12 weight % of soil particles passing a No. 200 (75-μm) sieve, depending upon the distribution of particle sizes which will cause them to have free-draining characteristics. This method utilizes vibratory compaction to obtain maximum density and pouring to obtain minimum density.

2. Definition

2.1 *relative density*—the state of compactness of a soil with respect to the loosest and densest states at which it can be placed by the laboratory procedures described in this method. It is expressed as the ratio of (*1*) the difference between the void ratio of a cohesionless soil in the loosest state and any given void ratio, to (*2*) the difference between its void ratios in the loosest and densest states. Mathematical expressions for this definition are presented in Section 8.

3. Apparatus

3.1 The assembly of the apparatus is shown in Fig. 1. Individual components and accessories shall be as follows:

3.1.1 *Vibratory Table*—A steel table with a cushioned steel vibrating deck about 30 by 30 in. (762 by 762 mm) actuated by an electromagnetic vibrator. The vibrator should be a seminoiseless type with a net weight over 100 lb (45.4 kg). The vibrator shall have a frequency of 3600 vibrations per minute, a vibrator amplitude variable between 0.002 and 0.025 in. (0.05 and 0.64 mm) under a 250-lbf (1112-N) load, and shall be suitable for use with 230-V ac.

3.1.2 *Molds*—Cylindrical metal unit-weight molds of 0.1 and 0.5 ft³ (2830 cm³ and 14160 cm³) capacity, conforming to the dimensional requirements as shown in Fig. 2.

3.1.3 *Guide Sleeves*—One guide sleeve with clamp assembly (see Fig. 3(*a*)) for each size mold. Two of the three set screws on the clamp assembly should be provided with lock nuts.

3.1.4 *Surcharge Base Plates*—One surcharge base plate ¹⁄₂ in. (12.7 mm) in thickness for each size mold.

3.1.5 *Surcharge Weights*—One surcharge weight (see Fig. 4) for each size mold. The total weight of surcharge base plate and surcharge weight shall be equivalent to 2 psi (14 kPa) for the mold being used.

3.1.6 *Surcharge Base Plate Handle*—One for each surcharge base plate (see Fig. 3(*b*)).

3.1.7 *Dial Indicator Gage Holder*, as shown in Fig. 3(*c*).

3.1.8 *Dial Indicator*, 2-in. (50.8-mm) travel with 0.001-in. (0.025-mm) graduations.

3.1.9 *Calibration Bar*, metal, 3 by 12 by ¹⁄₈ in. (76.2 by 304.8 by 3.2 mm).

---

[1] This method is under the jurisdiction of ASTM Committee D-18 on Soil and Rock.
Current edition approved Sept. 19, 1969. Originally issued 1964. Replaces D 2049 – 64 T.

[2] This method is currently under extensive revision by ASTM Committee D-18 because of the many limitations pointed out in *Evaluation of Relative Density and Its Role in Geotechnical Projects Involving Cohesionless Soils*, *ASTM STP 523*, Am. Soc. Testing Mats., 1973.

3.1.10 *Pouring Devices*—Pouring devices consisting of funnels ½ in. (12.7 mm) and 1 in. (25.4 mm) in diameter by 6 in. (152.4 mm) in length, with cylindrical spouts and lipped brims for attaching to 6 in. diameter by 12 in. (304.8 mm) high metal cans.

3.1.11 *Mixing Pans*—Two sizes of metal pans, one approximately 2 by 3 ft by 4 in. (0.6 by 0.9 m by 102 mm) deep and the other approximately 16 by 16 by 2 in. (406 by 406 by 51 mm) deep.

3.1.12 *Scale*—One portable platform scale, 250-lb (11.3-kg) capacity, with 0.01-lb (4.5-g) graduations.

3.1.13 *Hoist*—A rope, chain, or cable hoist of at least 300 lb (136-kg) capacity.

3.1.14 *Other Equipment*, including a large metal scoop, a hair-bristled dusting brush, a timing device indicating in minutes and seconds, a 15-in. (381-mm) metal straightedge, and a 0 to 1-in. (0 to 25-mm) micrometer accurate to 0.001 in. (0.025 mm).

4. Calibration

4.1 Determine the volume of the mold by direct measurement and check the volume by filling with water as provided in 4.1.1 or 4.1.2. Determine the initial dial reading for computing the volumes of the specimen as provided in 4.1.3.

4.1.1 *Volume by Direct Measurement*—Determine the average inside diameter and height of the mold to 0.001 in. (0.025 mm). Calculate the volume of the 0.1 ft³ (2830-cm³) mold to the nearest 0.0001 ft³ (2.83 cm³) and the 0.5-ft³ (14 160-cm³) mold to the nearest 0.001 ft³ (28.3 cm³). Calculate also the average inside cross-sectional area of the mold in square feet.

4.1.2 *Volume by Filling with Water*—Determine the weight of the water, in grams, required to fill the mold. Slide a glass plate carefully over the top surface of the mold in such a manner as to ensure that the mold is completely filled with water. Determine the temperature of the water in the mold. A thin film of cup grease on the top surface of the mold will make a watertight joint between the glass plate and top of the mold. Calculate the volume of the mold, in cubic feet, by multiplying the weight of water, in grams, used to fill the mold by the unit volume of water, in millilitres per gram at the observed tempera-ture taken from Table 1, and dividing the result by 28 317 ml/ft³.

4.1.3 *Initial Dial Reading*—Determine the thicknesses of the surcharge base plate and the calibration bar to 0.001 in. (0.025 mm) using a micrometer. Place the calibration bar across a diameter of the mold along the axis of the guide brackets. Insert the dial indicator gage holder in each of the guide brackets on the measure with the dial gage stem on top of the calibration bar and on the axis of the guide brackets. The dial gage holder should be placed in the same position in the guide brackets each time by means of matchmarks on the guide brackets and the holder. Obtain six dial indicator readings, three on the left side and three on the right side, and average these six readings. Compute the initial dial reading by adding together the surcharge base plate thickness and the average of the six dial indicator readings and subtract the thickness of the calibration bar. The initial dial reading is constant for a particular measure and surcharge base plate combination.

5. Sample

5.1 Select a representative sample of soil. The weight of sample required is determined by the maximum size of particle as prescribed in Table 2.

5.2 Dry the soil sample in an oven at a temperature of 230 ± 9 F (110 ± 5 C). Process the soil through a sieve with openings sufficiently small to break up all weakly cemented soil particles.

6. Minimum Density Procedure

6.1 Determine the minimum density (zero relative density), (maximum void ratio) as follows:

6.1.1 Select the pouring device and mold according to the maximum size of particle as indicated in Section 5. Weigh the mold and record the weight. Use oven-dried soil.

6.1.2 Place soil containing particles smaller than ³⁄₈ in. (9.5 mm) as loosely as possible in the mold by pouring the soil from the spout in a steady stream while at the same time adjusting the height of the spout so that the free fall of the soil is 1 in. (25.4 mm). At the same time, move the pouring device in a spiral motion from the outside toward the center to form a soil layer of uniform thickness without segregation. Fill the mold approximately 1 in. above the top and screed off the excess soil level with the top by making one continuous pass with the steel straightedge. If all excess material is not removed, an additional continuous pass shall be made but great care must be exercised during the entire pouring and trimming operation to avoid jarring the mold.

6.1.3 Place soil containing particles larger than $^3/8$ in. (9.5 mm) by means of a large scoop (or shovel) held as close as possible to and just above the soil surface to cause the material to slide rather than fall onto the previously placed soil. If necessary, hold large particles back by hand to prevent them from rolling off the scoop. Fill the mold to overflowing but not more than 1 in. (25.4 mm) above the top. With the use of the steel straightedge (and the fingers when needed), level the surface of the soil with the top of the measure in such a way that any slight projections of the larger particles above the top of the mold shall approximately balance the larger voids in the surface below the top of the mold.

6.1.4 Weigh the mold and soil and record the weight.

7. Maximum Density Procedure

7.1 Determine the maximum density (100% relative density, minimum void ratio) by either the dry or wet method as follows:

7.1.1 *Dry Method:*

7.1.1.1 Mix the sample of oven-dried soil to provide an even distribution of particle sizes with as little segregation as possible.

7.1.1.2 Assemble the guide sleeve on top of the mold and tighten the clamp assemblies so that the inner wall of the sleeve is in line with the inner wall of the mold. Tighten the lock nuts on the two set screws equipped with lock nuts. Loosen the clamp assembly having no lock nuts, remove the guide sleeve. Weigh the empty mold and record the weight.

7.1.1.3 Fill the mold with soil by the procedure specified in 6.2 or 6.3.

NOTE 1—Normally, the mold filled with soil for the minimum density determination may be used for the maximum density determination without refilling the mold.

7.1.1.4 Attach the guide sleeve to the mold and place the surcharge base plate on the soil surface. Lower the surcharge weight onto the

surcharge base plate, using the hoist in the case of the 0.5-ft^3 (14 160 cm^3) mold.

7.1.1.5 Set the vibrator control at maximum amplitude and vibrate the loaded specimen for 8 min. Remove the surcharge weight and guide sleeve from the mold. Obtain and record dial indicator gage readings on two opposite sides of the surcharge base plate average, and record the average. Weigh the mold and soil if this has not been done in the minimum density determination or if an appreciable amount of fines has been lost during vibration. Record the weight.

7.1.2 *Wet Method:*

NOTE 2—While the dry method is preferred from the standpoint of securing results in a shorter period of time, the highest maximum density is obtained for some soils in a saturated state. At the beginning of a laboratory testing program, or when a radical change of materials occurs, the maximum density test should be performed on both wet and dry soil to determine which method results in the higher maximum density. If the wet method produces higher maximum densities, (in excess of one percent) it shall be followed in succeeding tests.

7.1.2.1 The wet method may be conducted on oven-dried soil to which sufficient water is added or, if preferred, on wet soil from the field. If water is added to dry soil allow a minimum soaking period of $^1/2$ h.

7.1.2.2 Fill the mold with wet soil by means of a scoop or shovel. Add sufficient water to the soil to allow a small amount of free water to accumulate on the surface of the soil during filling. The correct amount of water can be estimated by a computation of the void ratio at expected maximum density or by experimentation with the soil. During and just after filling the mold, vibrate the soil for a total of 6 min. During this period reduce the amplitude of the vibrator as much as necessary to avoid excessive boiling and fluffing of the soil, which may occur in some soils. During the final minutes of vibration, remove any water appearing above the surface of the soil.

7.1.2.3 Assemble the guide sleeve, surcharge base plate, and surcharge weight as described in 4.1.1.

7.1.2.4 Vibrate the specimen and surcharge weight for 8 min. After the vibration period, remove the surcharge weight and guide sleeve from the mold. Obtain and record dial indicator gage readings on two opposite sides of the surcharge base plate. Carefully remove the entire wet specimen from the mold and dry to

D 2049

constant weight. Weigh dry specimen and record.

8. Calculations

8.1 *Minimum Density*—Calculate minimum density, γ_{min}, in pounds per cubic foot, as follows:

$$\gamma_{min} = W_s/V_c$$

8.2 *Maximum Density*—Calculate maximum density, γ_{max}, in pounds per cubic foot, as follows:

$$\gamma_{max} = W_s/V_c$$

where:

W_s = weight of dry soil, lb,
V_c = calibrated volume of mold, ft³,
V = volume of soil, ft³ = $V_c - (R_i - R_f)/12 \times A$,
R_f = final dial gage reading on the surcharge base plate after completion of the vibration period, in.,
R_i = initial dial gage reading, in., and
A = cross-sectional area of mold, ft².

8.3 *Density of Soil in Place*—Determine the density of the soil in place, γ_d, in a compacted fill or a natural deposit in accordance with either ASTM Method D 1556, Test for Density of Soil in Place by the Sand-Cone Method[3] or ASTM Method D 2167, Test for Density of Soil in Place by the Rubber-Bal-

loon Method.[3]

8.4 *Relative Density*—Calculate relative density, D_d, expressed as a percentage as follows:

$$D_d = [\gamma_{max}(\gamma - \gamma_{min})/\gamma(\gamma_{max} - \gamma_{min})] \times 100$$

or in terms of void ratio

$$D_d = [(e_{max} - e)/(e_{max} - e_{min})] \times 100$$

where:

e = volume of voids divided by the volume of solid particle,
e_{max} = void ratio in loosest state, and
e_{min} = void ratio in most compact state.

9. Report

9.1 The report shall include the following information. A recommended form is shown in Fig. 5.

9.1.1 Method used for determining minimum density (funnel or scoop),

9.1.2 Method used for obtaining maximum density (dry or wet, and mold size),

9.1.3 Minimum and maximum densities,

9.1.4 Density "in-place" of the field sample, and

9.1.5 Relative density of the field sample.

---

[3] *Annual Book of ASTM Standards*, Part 19.

This standard is subject to revision at any time by the responsible technical committee and must be reviewed every five years and if not revised, either reapproved or withdrawn. Your comments are invited either for revision of this standard or for additional standards and should be addressed to ASTM Headquarters. Your comments will receive careful consideration at a meeting of the responsible technical committee, which you may attend. If you feel that your comments have not received a fair hearing you should make your views known to the ASTM Committee on Standards, 1916 Race St., Philadelphia, Pa. 19103, which will schedule a further hearing regarding your comments. Failing satisfaction there, you may appeal to the ASTM Board of Directors.

TABLE 1   Volume of Water per Gram Based on Temperature[a]

Temperature		Volume of Water,
deg C	deg F	ml/g
12	53.6	1.00048
14	57.2	1.00073
16	60.8	1.00103
18	64.4	1.00138
20	68.0	1.00177
22	71.6	1.00221
24	75.2	1.00268
26	78.8	1.00320
28	82.4	1.00375
30	86.0	1.00435
32	89.6	1.00497

[a] Values other than shown may be obtained by referring to the *Handbook of Chemistry & Physics*, Chemical Rubber Publishing Co., Cleveland, Ohio.

ASTM **D 2049**

TABLE 2 **Required Weight of Sample.**

Maximum Size of Soil Particle in. (mm)	Weight of Sample Required, lb (kg)	Pouring Device To Be Used in Minimum Density Test	Size of Mold To Be Used, ft³ (cm³)
3 (76.2)	100 (45.4)	Shovel or extra large scoop	0.5 (14 160)
1½ (38.1)	25 (11.3)	Scoop	0.1 (2 830)
¾ (19.1)	25 (11.3)	Scoop	0.1 (2 830)
⅜ (9.5)	25 (11.3)	Pouring Device (1-in. (25.4-mm) dia. spout)	0.1 (2 830)
No. 4 (4.75)	25 (11.3)	Pouring Device (½-in. (12.7-mm) dia. spout)	0.1 (2 830)

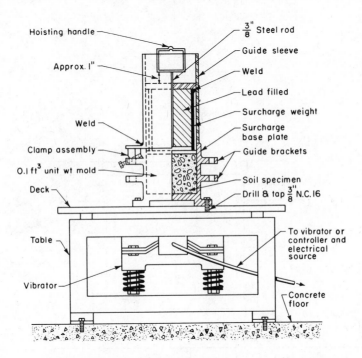

FIG. 1 **Assembly of Apparatus (with 0.1 ft³ (2830 cm³) Mold Assembly).**

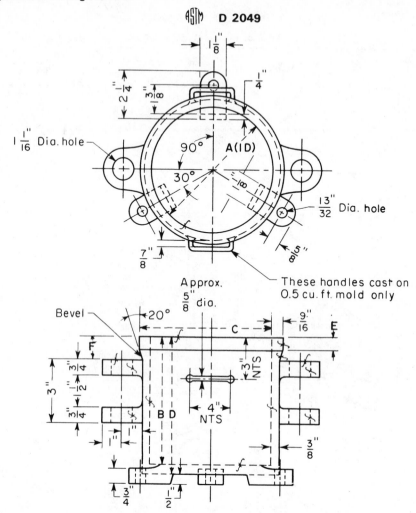

FIG. 2   Details of Molds.

Size Mold ft³ (cm³)	Dimensions, in. (mm)					
	A	B	C	D	E	F
0.1 (14 160)	6.000 (152.4)	6.112 (155.19)	7¹⁄₈ (181.0)	6¹⁄₂ (105.1)	¹⁄₂ (12.7)	1¹⁄₈ (28.6)
0.5 (2 830)	11.000 (279.4)	9.092 (230.89)	12¹⁄₈ (308.0)	9¹⁄₂ (241.3)	⁵⁄₈ (15.9)	2 (50.8)

ASTM D 2049

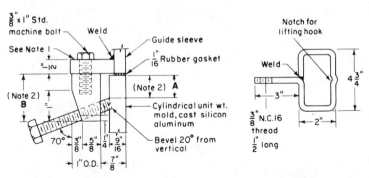

(a)
Clamp Assembly and Guide Sleeve for Guided
Surcharge Weights.

(b)
Surcharge Base Plate Handle.
(I Required)

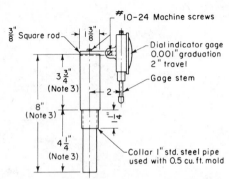

(c)
Holder for Dial Indicator Gage

NOTE 1—This piece shall be a steel bar, 1½ by ½ in. (38.1 by 12.7 mm) of a length necessary to produce the indicated dimension from the inside of the guide sleeve. Weld 3 clamp assemblies to the guide sleeve at equal spacing.

NOTE 2—See the following tabulation.

Size Mold, ft³ (cm³)	A, in. (mm)	B, in. (mm)	Guide Sleeve, in. (mm)
0.1 (14,160 cm³)	½ (12.7)	1⅜ (34.9)	Steel tubing, 6 in. (152.4 mm) ID, ¼-in. (6.4-mm) wall, 12 in. long (304.8 mm)
0.5 (2,830 cm³)	⅝ (15.9)	1½ (38.1)	Steel pipe, 11 in. (279.4 mm) ID, ⅜-in. (9.5-mm) wall, 8 in. (203.2 mm) long

NOTE 3—These dimensions must be changed to fit the dial gage indicator used.

FIG. 3 Details of Apparatus Components.

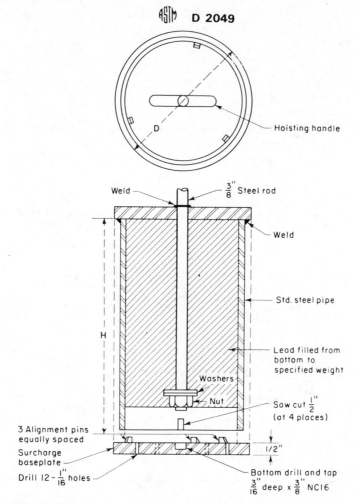

ASTM   **D 2049**

Size Mold, ft³ (cm³)	D, in. (mm)	H, in. (mm)	Std. pipe, in. (mm)	Total wt. required, lb (kg)
0.1 (14 160)	5¹⁵⁄₁₆ (150.8)	9 (228.6)	4 (101.6)	57 ± 0.5 (25.9 ± 0.2)
0.5 (2 830)	10⅞ (276.2)	6 (152.4)	10 (254.0)	190 ± 1.0 (8.6 ± 0.4)

NOTE 1—All plates shall be ¹₂-in. (12.7-mm) thick steel.

NOTE 2—Top plates for weights may be torch-cut, but edges must be ground as smooth as practicable. Surcharge base plates must be machined to the specified diameter.

NOTE 3—Hoisting handles shall have the same shape as the Surcharge Base Plate Handle (see Fig. 3(b)).

FIG. 4   Surcharge Weight.

ⒶⓈⓉⓂ D 2049

RELATIVE DENSITY DETERMINATIONS

Project_____ Feature_____ Sample No._____

Tested by_____ Computed by_____ Checked by_____

Minimum Density Determination (0% Relative Density)			
Test No.			
Wt. soil + mold lbs.			
Wt. mold lbs.			
Wt. soil (W_s) lbs.			
Volume of mold (V_c) cu.ft.			
Minimum Dens. $= \dfrac{W_s}{V_c}$ pcf			

Relative Density Computation			
Test No.			
① In-place density pcf			
② Max. lab. density pcf			
③ Min. lab. density pcf			
④ ① - ③			
⑤ ④ x ②			
⑥ ② - ③			
⑦ ① x ⑥			
Relative Dens. $\% = \dfrac{⑤}{⑦} \times 100$			

Maximum Density Determination (100% Relative Density)	Dry Method	Wet Method
Test No.		
Left gage read. inches		
Right gage read. inches		
Avg. gage read. R_f		
Initial gage read. R_1		
Area of sample surface sq.ft. A		
Calib. vol. of mold cu.ft. V_c		
Soil vol. = V_c - $\dfrac{R_1 - R_f}{12}$ x A V_s		
Wt. dry soil + mold lbs.		
Wt. mold lbs.		
Wt. dry soil lbs. W_s		
Maximum Density $\dfrac{W_s}{V_s}$ pcf		

Mold No._____Surcharge base plate no._____
Surcharge base pl. thick._____in.
Straightedge thickness_____in.
Left dial read._____ _____ _____
Right dial read._____ _____ _____
R_1 = Avg. dial gage read. + surcharge base pl. thick. - st.edge thick.
R_1 = _____in.

FIG. 5 Recommended Report Form.

GLOSSARY OF TERMS

Conventional units in the English and metric systems are shown with terms where they apply. Where no units are shown, the terms are dimensionless.

B	footing width (ft)
C_c	coefficient of curvature
C_c	compression index
C_u	uniformity coefficient
D_f	depth of footing (ft)
D_n	maximum particle size in finer nth percentage (mm)
D_r	relative density (%)
D_{10}	effective size of soil particles (mm)
FS	factor of safety
FS_c	factor of safety with respect to cohesion
FS_ϕ	factor of safety with respect to friction
G_s	specific gravity
H	longest path taken by water seeping from the soils as a result of application of the consolidating pressure increment, Δp (in., cm, or ft)
H	height of water surface above datum (ft)
H_c	critical height of slope (ft)
H_w	height of water surface in well (ft)
K	coefficient of lateral earth pressure
K_A	active earth pressure coefficient
K_o	coefficient of earth pressure at rest
LL	liquid limit (%)
N_c	bearing capacity factor for cohesion
N_q	bearing capacity factor for surcharge
N_s	stability number
N_γ	bearing capacity factor for soil weight
P	wheel load (kips, lb)
PI	plasticity index (%)
PL	plastic limit (%)
Q	total flow volume (ft^3)

Q_d	ultimate load capacity of deep foundation (tons, kips)
Q_p	point load capacity of deep foundation (tons, kips)
Q_s	friction load capacity of deep foundation (tons, kips)
R	radius of influence of well (ft)
R_w	radius of well (ft)
SL	shrinkage limit (%)
S_R	degree of saturation (%)
T_s	surface tension (g/cm)
T_u	time factor for consolidation analysis
V	total soil volume (cm^3 or ft^3)
V_A	volume of air in soil (cm^3 or ft^3)
V_S	solid volume in soil (cm^3 or ft^3)
V_V	total soil void volume (cm^3 or ft^3)
V_W	volume of water in soil (cm^3 or ft^3)
W	pile hammer ram weight (kips, lb)
W	total weight of soil volume (g or lb)
W_S	weight of soil solids (g or lb)
W_W	weight of water in soil (g or lb)

c	cohesion (tsf, ksf, or psf)
c	constant for *Engineering News* pile driving formula (in.)
c_c	critical cohesion (tsf, ksf, or psf)
c_v	coefficient of consolidation (cm^2/sec)
$c_{req'd}$	cohesion required for stable slope (tsf, ksf, or psf)
d	diameter of capillary (cm)
d_p	diameter of tire contact area (in.)
d_s	width of loaded area on subgrade (ft, in.)
e	void ratio
e_1	initial void ratio
$e_{(max)}$	maximum void ratio determined by standard test method
$e_{(min)}$	minimum void ratio determined by standard test method
f_s	unit side friction for deep foundation (tsf, ksf, or psf)
h	height of drop of pile hammer (ft)
h_1	thickness of compressible stratum (ft)
i	hydraulic gradient
k	permeability coefficient (cm/sec)
n	porosity (%)
p	total stress (ksf or psf)
p	tire pressure (psi)
p'	effective stress (ksf or psf)
p_0	vertical effective stress (ksf or psf)
p_1'	initial effective stress (ksf or psf)
p_2'	effective stress after increment, Δp (ksf or psf)
q	flow rate (water; cfs)

q_a allowable bearing pressure (ksf or psf)

q_d ultimate bearing capacity (ksf or psf)

q_p net unit point resistance for deep foundation (tsf, ksf, or psf)

q_p tire contact stress on pavement (psi)

q_s pressure on subgrade (psf)

q_u unconfined compressive strength (tsf, ksf, or pcf)

s penetration of the pile resulting from one hammer blow (in.)

s settlement (in.)

s shearing strength (tsf, ksf, or psf)

t pavement thickness (in.)

t_u time for a given percentage of consolidation to occur (min, mo, or yr)

u porewater pressure (ksf or psf)

v apparent velocity of flow (fps)

w water content (%)

z_c height of capillary rise (cm or ft)

α unslope angle (°)

α meniscus contact angle (°)

β slope angle from the horizontal (°)

γ density or unit weight (pcf)

γ_d dry density or dry unit weight (pcf)

$\gamma_{d(max)}$ maximum dry density determined by standard test method (pcf)

$\gamma_{d(min)}$ minimum dry density determined by standard test method (pcf)

$\gamma_{d(zav)}$ dry density for saturated soil at specified water content (pcf)

γ_s density or unit weight of solid material (pcf)

γ_w density or unit weight of water (pcf)

Δh settlement (in.)

Δp effective stress increment (ksf or psf)

ϕ angle of internal friction (°)

$\phi_{req'd}$ angle of internal friction required for stable slope (°)

ϕ_s' approximate angle of internal friction for slope with seepage (°)

Unified Classification Symbols

G   gravel

S   sand

M   silt

C   clay

O   organic

Pt   peat

W   well graded soil

P   poorly graded soil

L   low compressibility

H   high compressibility

NP   nonplastic

PROBLEM SOLUTIONS

Chapter 3

1 (a) 113.6 lb/ft^3
  (b) 100 lb/ft^3
  (c) 0.63
  (d) 39 percent
  (e) 59 percent
  (f) 124 lb/ft^3

2 (a) 105.5 lb/ft^3
  (b) 68.6 lb/ft^3

3 0.26 tons

4 (a) 103.4 lb/ft^3
  (b) 37.2 percent
  (c) 0.592
  (d) 71.5 percent

5 114 percent

6 20 gal/yd^3

7 104.9 lb/ft^3, 73.07 lb/ft^3, 2.38

Chapter 4

1 OL

2 SM

3

Soil No.	Unified	AASHTO
2	SC	A-2-6 (1)
4	MH	A-7-5 (17)
6	CL	A-6 (17)

Chapter 5

1

Depth (ft)	Total, p_v (lb/ft²)	Water, u (lb/ft²)	Effective, p_v' (lb/ft²)
0 (surface)	0	0	0
10 (water table)	1100	0	1100
32 (stratum change)	3828	1373	2455
50 (bedrock)	5628	2496	3132

2 624 psf

3 49 ft

4 $k = 1.0 \times 10^{-2}$ cm/sec

5 $k = 3.0 \times 10^{-2}$ cm/sec

6 Improve by including an observation well.

7 $C_c = 0.63$ and $C_c' = 0.09$

8 $c_v = 1.2 \times 10^{-4}$ cm²/sec

9 Undisturbed $q_u = 0.50$ tsf and remolded $q_u = 0.30$ tsf

10 $\phi = 34°$

11 $q_u = 345$ psf

Chapter 8

1 89.9 pcf, 27.8 percent

2 An error was made in the test.

4 $\gamma_d = \dfrac{\gamma_w}{w + \dfrac{1}{G_s}}$

5 Specifications not met.

6 (a) $D_r = 59$ percent
   (b) $D_r = -9$ percent
   (c) $D_r = 117$ percent

7 15 percent

8 $\gamma_{d-\#4} = 124.1 \ \text{lb/ft}^3$
$w_{-\#4} = 7.9 \ \text{percent}$

Chapter 9

1 (a) Dewatering method: Wellpoints; initial spacing is 12 ft.
  (b) Dewatering method: Cofferdam.

2 (a) Dewatering method: Sump pump.
  (b) Dewatering method: Caisson.
  (c) Do not plan operation without consultation of specialty subcontractor.

Chapter 10

1 FS = 3.4. Cut can be made safely, and normal work can progress without danger.

2 Trench will have to be structurally supported.

3

Depth (ft)	p'_v (lb/ft²)	p'_h (lb/ft²)	u (lb/ft²)	p_h (lb/ft²)
0	0	0	0	0
10	1100	440–550	0	440–550
32	2455	982–1228	1373	2355–2601
50	3132	1253–1566	2496	3749–4062

5 Support method: Soldier piles and horizontal lagging.

6 Support method: Vertical sheeting with tie backs.

Chapter 11

1 For 5 ft footing, estimated settlement = 1.72 in.
For 10 ft footing, estimated settlement = 2.42 in.

2 FS = 4.29.

3 $P = 21$ kips.

4 $P = 15$ kips.

5 Hammer selections: Vulcans 50C, 65C, or 80C. 80C. $Q_d = 108$ kips

6 Length estimate $= 80$ ft.
Hammers: Vulcan 80C or 140C.

7 Select a single-acting hammer with energy of 30,000 to 40,000 ft-lb, such as MKT S-10, Conmaco 100, Raymond 210, or MKT S-14.

8 Static capacity calculations show that you should continue driving piling. Consider a load test to demonstrate acceptance.

9 Belled piers at 26 ft.

Chapter 12

1 46 in.

2 Compaction improves capacity 2.5 to 3 times.

3 Yes; FS $= 1.47$.

4 Compacted sand will not work. Requires 18-in. cement-stabilized sand.

5 $t = 34$ in. without geotextile; 19 in. with geotextile.

INDEX

325